THIS IS
YOUR CAPTAIN
SPEAKING

THIS IS
YOUR CAPTAIN
SPEAKING

Stories from the Flight Deck

DOUG MORRIS

Copyright © Doug Morris, 2022

Published by ECW Press
665 Gerrard Street East
Toronto, Ontario, Canada M4M 1Y2
416-694-3348 / info@ecwpress.com

Cover design: David Drummond

What I wrote and opined does not necessarily reflect the opinion of the airline I fly for. In that respect, I am flying solo. Rest assured, I obtained permission from my company's hierarchy before this book materialized.

This note is the small print comparable to that found on television car and pharmaceutical ads or depicted on paperwork for things you purchase, sell, or anything else in life involving a lawyer.

LIBRARY AND ARCHIVES CANADA CATALOGUING
IN PUBLICATION

Title: This is your captain speaking : stories from the flight deck / Doug Morris.

Names: Morris, Doug, author.

Description: Sequel to: From the flight deck: plane talk and sky science.

Identifiers: Canadiana (print) 20210340819 | Canadiana (ebook) 20210340886

ISBN 978-1-77041-585-0 (softcover)
ISBN 978-1-77305-797-2 (ePub)
ISBN 978-1-77305-798-9 (PDF)
ISBN 978-1-77305-799-6 (Kindle)

Subjects: LCSH: Aeronautics, Commercial—Anecdotes. | LCSH: Aeronautics—Anecdotes. | LCSH: Air travel—Anecdotes. | LCSH: Airplanes—Piloting—Anecdotes. | LCSH: Airplanes—Anecdotes.

Classification: LCC TL720 .M675 2022 | DDC 629.13—dc23

This book is funded in part by the Government of Canada. *Ce livre est financé en partie par le gouvernement du Canada.* We also acknowledge the support of the Government of Ontario through the Ontario Book Publishing Tax Credit, and through Ontario Creates.

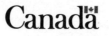

PRINTED AND BOUND IN CANADA

PRINTING: MARQUIS   5 4 3 2 1

I dedicate this book to two young aspiring pilots,
Alex McIntyre and Jared Logan,
both tragically killed while chasing their dreams,
February 17, 2020.

✈

THE FLIGHT PLAN

CHAPTER 6: PRE-DESCENT, IN-RANGE, LANDING AND AFTER LANDING CHECKLIST

I drew a picture of a pair of wings . . . because I want to fly.
My mother asked me to explain . . . I said that I would try.
I had a dream the other night about flying . . .

— "DREAM NO. 2," KEN TOBIAS

(One of my favorite songs)

THIS IS YOUR CAPTAIN SPEAKING . . .

The flight deck door is incessantly closed. Sure, you get to see pilot's approaches, landings, and takeoffs spewed over the internet, but dialogue with a flight crew is not easy. Passengers are intimidated to approach a pilot and ask questions, thinking there may be repercussions. The only ones without inhibitions are kids.

✈ Aviation often dates the Wright brothers' first flight to 1903, but modern-day aviation can be divided into pre-9/11 and post-9/11, akin to the Julian calendar dividing time into BC and AD. We have now surpassed 9/11's 20th anniversary, and yet the aftermath still plagues aviation. One downfall is the dissemination of the intricacies of aviation.

✈ One positive spin-off from 9/11 is the record-setting musical *Come from Away*, which narrates what happened when more than 30 airliners commandeered the town of Gander, Newfoundland, after the attacks. I attended the show in Toronto. It sure induced teary eyes because it hit home: I was flying the day aviation went inverted.

I hope to fill the void of aviation knowledge by enlightening a wide spectrum of readers. Sort of what my in-flight magazine articles have accomplished, but there will be no one in the background taming my thoughts with a torrent of editing. Now 24 years and counting, I have written for the occasionally flown passenger, but these articles had to be dumbed down and politically correct. But wisdom will still prevail here; you won't get *all* the aviation dirt from this book.

If you are curious about the "nuts and bolts" of aviation, or if you are a spouse, relative, friend, or an inquisitive neighbor to an aviator, this book is for you! And if you have a fear of flying, the aviation intricacies you'll learn will lessen your anxiety.

My aviation career is on its final approach. It's been a great run, so before I set the parking brake for the last time, I feel compelled to write the sequel to my first book, *From the Flight Deck: Plane Talk and Sky Science*. It too had a fantastic run.

ABOUT ME.

Whenever anyone presents a speech, writes a book, or teaches a class, there is usually a short intro about who the author or presenter is. Here's mine.

Presently, I fly a 298-passenger Boeing B787 (Dreamliner) around the world for an airline with a maple leaf emblazoned on its fuselage; I have amassed over 26,000 hours of total flight time. To give perspective on this total time, it's equivalent to driving from Toronto to Montreal (or Boston to New York, or Los Angeles to San Francisco) and back again daily for nearly seven years. When I write this analogy, I too am shocked how much time I've flown. And this excludes flying during vacations, deadheading, and commuting — more about these last two later.

> ✈ To liven up my weather classes, I would tell this corny quip to the new-hire pilots: How does a flight attendant know their date with a pilot is halfway over? It's when the pilot stops and says, "Enough about me, let's talk about you . . . what do you think about me?" I've told that same joke over 150 times.

I am also a certified meteorologist, having worked for Environment Canada for four years as a forecaster, mostly on Canada's east coast. Again, I have been writing the aviation column for in-flight magazine *enRoute* and have written many aviation articles for various magazines and papers. The name Doug Morris has been in print over 300 times. This is my fourth book. *From the Flight Deck: Plane Talk and Sky Science* came first, followed by two aviation weather books that took several years to write. But enough about me, let's talk about you. What do you think about me?

Most pilots, including me, were bitten by the aviation bug early in life. It's a disease only curable by learning to fly. Over my career, I can't count how many times I've heard, "I always wanted to be a pilot." I felt like an aviation priest listening to confessions, with most coming from complete strangers.

> ✈ I guess you can say I have also been bitten by a latent writing bug. I didn't think I would be saying this, especially having been jinxed by high school and university teachers and some editors averring that writing would not be a viable hobby. Glad I proved them wrong.
>
> ✈ Yet another bug is the tenacious travel bug. My three kids inherited it as well. My youngest has been to 50 countries.

CHAPTER 1

PLANNING — BEFORE THE FLIGHT

MY ZIG-ZAG APPROACH IN BECOMING AN AIRLINE PILOT.

Growing up on Canada's east coast, I was generally removed from the pulse of aviation. Sure, a small airline called Eastern Provincial Airlines dominated, but they were swallowed up by their Upper Canada counterparts during mergers. Knowing a pilot equated to knowing an astronaut, and seeking wisdom from high school guidance counselors proved futile. It is why, to this very day, I will always take time to answer questions and to mentor those pining for the skies. My mother sensed my quest for the skies and bought a familiarization flight for my 16th birthday. She even joined me on the flight. She passed away early in life, but before she went, she had a dream of me wearing a dark blue suit. We did have a neighbor down the street who became a military pilot. I shyly approached him one winter's day while playing hockey on a local lake, but let's just say PR skills were not his asset. Another reason why I stop what I am doing and bestow wisdom to future pilots.

Science courses would take me further (so I thought) than the arts, and that's what I went after, concentrating on physics, chemistry, and math and thumbing my nose at the artsy English majors. I took

French, thinking it might be an additional stepping stone in aviation. Painting houses on the east coast and planting trees on Canada's west coast, I earned enough money for flying lessons at the Halifax Flying Club in Nova Scotia. Sadly, that club and other nearby clubs no longer exist in the largest city on the east coast. After a three-year degree in physics, and a fresh commercial pilot license from Gimli, Manitoba, my dream stalled when I was 21. A recession plagued the economic scene. Time to head back to university. Aviation is one of the first industries to feel the plight of a recession, and one of the last to recover. I gave the military a try, but they weren't hiring. I studied meteorology at McGill University in Montreal and then went to Toronto to become a certified meteorologist for Atmospheric Environment Canada. But creating wave height forecasts as a civilian forecaster for the military during the wee hours had me looking out the office window at three a.m., thinking there must be something better. A stint as a weather guy in Esquimalt, B.C., had me briefing the admiral in the morning and taking advanced flying lessons in the afternoon. I returned to the east coast with my bare minimum qualifications to fly as a commercial pilot.

When I flew home with my new multi-engine IFR (Instrument Flight Rules) ratings in my wallet, our national airline was on strike, so I scrambled to take the competition — a Canadian Airlines flight. As I settled into my seat, an agent approached me explaining my seat was "duped," airline talk meaning "duplicated" — as in, I had to deplane. I'm not sure what possessed me, but I quickly mentioned I had my commercial pilot license and wondered if I could sit in the jumpseat. That fateful act whisked me to the flight deck with my license in hand. I'll never forget how, landing in Halifax in light snow showers and low cloud on a dark, wintery night, the captain casually put his cigarette out and flew the non-precision approach. That proved to be a pinnacle. Days later, I made a phone call to a local air cargo company flying bank bags and was hired part-time flying small twin-engine

Navajos. After a year, I threw in the towel as a weatherman and pursued aviation full-time. I flew for Air Atlantic, then wing-walked to Air Nova, paving the way to my present airline, where I started as a 34-year-old cruise pilot flying the now-mothballed four-engine Airbus A340. But that is my story, and every pilot has theirs. In fact, everyone has a story.

WHY BECOME A PILOT AND WHAT'S SO GREAT ABOUT IT?

The short answer to why being a pilot is so great: because the job is dynamic. That's my quick response when asked. But there are a multitude of pluses. Firstly, it brings travel, and travel magnifies life. I've seen so many people around the world in mundane jobs mired in the rut of life. I recently heard that nearly 75 percent of people dislike their job. As an airline pilot, you rarely have the same day. Weather is tops in inducing variability. It's also a very respected job and gives your kids bragging rights, as many kids don't even know what their parents do. Aviating comes with a ton of perks but requires skill, the ability to think on your feet, flexibility, well-roundedness, and generally a type A personality. Sure, there are drawbacks, but the pros far outweigh the cons. I know it's where I want to be, so I thank my lucky stars it all worked out.

✈ It's not only the flying, but the comradery and sharing the aviation passion that make it the best job in the world. I have met retired pilots on layovers that flew in for the night just to absorb the vibe of yesteryear and reminisce about the good ole days. Sure, some retired pilots never look back, but the majority hold their flying careers close to their hearts.

> ✈ Nowhere will you witness more of the passion for aviation than at the Experimental Aircraft Association's annual EAA AirVenture Oshkosh air show in Wisconsin — truly the world's most famous aviation mecca. There, this unassuming airport hosts nearly 600,000 visitors in slightly over a week. But be forewarned: don't bring a non-aviator to this event. However, if you want to see where a huge proportion of men over 50 has convened, then look no further.

Waiting for a North American major airline to call is no longer the sole solution, as you can be hired in most countries, but there are some drawbacks to flying abroad. Being an expat pilot has its pitfalls and is not for everyone. Overseas airlines are using you for your license and experience, and any hiccup in some countries means you are heading home in a hurry or held against your will. But seeing and living in another culture, let alone the fantastic pay with great tax savings, can take the sting out of things.

> ✈ I too pondered a lavish overseas contract. I mean, who wouldn't mind escaping hefty income tax payments? I took my family to Dubai during my interview with Emirates Airlines. Though they promised impeccable working conditions and benefits, I couldn't see the long-term benefit of living in a super-hot "sandbox." I know many that loved it, but others trickled back home with or without their marriage intact. It is there I learned these words of wisdom: "Happy wife, happy life."

Each new pilot can cite a unique career path. Until recently, pilots chased their dreams by rushing to the first company hiring. Each résumé depicting their journey is different, but now the trek has shortened and is more direct. Many smaller airlines are implementing "flow-through" programs. Jazz Aviation has Jazz Aviation Pathways Program, making the path both more direct and generic. This too is happening in the U.S., through programs referred to as gateways, pilot pathways, and career path programs, to name a few. Many of the major carriers are hiring directly from their connectors, and the connectors are hiring directly from flight schools, with many companies giving initiatives like sign-on bonuses. You don't have to work on the ramp up north in Tuktoyaktuk or Alaska for six months to a year, drooling over the chance of flying the airplane you are loading with oil drums — unless you want to.

The three general routes in becoming an airline pilot are through flying clubs, flight colleges/universities, and the road less traveled, the military. The Canadian military is an option for a select few mainly because of its size and needs. It has about 250 total aircraft in service, with helicopters making up nearly 60 percent. Something to think about if your goal is the airlines. They are certainly recruiting and have made it easier to apply, but they adhere to strict guidelines, and there is a good chance you will be flying a helicopter. In the USA, one has the air force, navy, army, marines, coast guard, national guard, etc., and they too come with stringent time commitments.

✈ Don't get me wrong about becoming a helicopter pilot. They too are much needed. I'm presently in downtown Los Angeles, visiting my daughter, who was hired to change the skyscape of the city core. For decades, it was mandatory that any tall building had to have a helipad in case of fire or other emergency. This flat-topped requirement has recently been overturned, but you'll notice an orange wind-sock bouncing about in the prevailing west wind on many tall buildings. In São Paulo, civilian helicopter pilots fly executive types above the city, safe from crime. I hear and feel their pulsating, whirling rotors during layover rests.

I've always averred flying for Canada's national airline or any major carrier in the United States is like making the cut for a professional sport team. You'll need a minimum of 2000 flight hours (this minimum varies and is falling), an ATPL (Airline Transport Pilot License), citizenship or landed immigrant status, and a category one medical. Points based on education, aircraft flown, experience, and languages are allotted when you apply online for the large carriers. As of now, the big American carriers are still asking for a degree, but that will falter as the pilot pool shrinks.

I pursued the flying-club and university route. I thought university would be a great stepping stone, plus it would allow me to fall back on something in case things didn't work out. I constantly get asked: What is the best route, flight school, college/university, or military?

And which school is the best? I tell budding pilots it must be their choice. Many want the fastest way to the airlines to get that almighty seniority number; thus I am asked how to build hours in the shortest time. This irks me a little, because they think they can put an old head on new shoulders, but I probably would have asked the same. Most want that "experience pill" giving them a university-level degree and thousands of hours of jet time at age 20, along with the right to walk into the chief pilot's office saying, "Hire me first!" As well, most don't realize there is little networking in the airlines, unlike in other fields. I can't walk into the office and throw someone's résumé on the chief pilot's desk and curtly say, "Hire this pilot!" It's all done online, although some companies allow a pilot to recommend one pilot per year. Every time I teach or present, I get Facebook and LinkedIn requests from pilot prospects. There is little I can do. But again, in their shoes, I would be doing the same thing. Everyone is taught that pestering, perseverance, and email nagging work — until they backfire. Yes, I've seen that happen.

GETTING YOUR CAREER AIRBORNE:
BECOMING A PILOT — FACTS AND URBAN MYTHS BUSTED.

People who need glasses can still become pilots. But having good hearing is just as important as good eyesight. I know of two pilots with one eye. I've flown with several diabetics, and a few pilots have hearing aids. Keep in mind, most of these pilots held aviation medicals before their medical plight. You don't need university, but it helps. Nor do you have to be bilingual in Canada — again it helps.

Starting an aviation career is just a phone call away to the nearest flight school. You can be up flying the same day, taking a "fam" (familiarization) flight. It's that easy! Some flying clubs call it a "discovery flight." If you think it may be a tad overwhelming, take a friend or two along, as many of these aircraft seat four. Once you become hooked after the "fam" flight — some will realize it's not for them — I recommend getting a pilot medical rather quickly. All flying clubs have a list of approved doctors. You don't want any medical hurdles, and if there are, leave no stone unturned if flight is your destiny. The biggest medical hiccup is that nearly 8 percent of males (0.5 percent of females) discover they may have a color vision deficiency. But don't let that stop you. There are tests that can determine your level of color deficiency. A longtime friend of mine was told the same thing, but he persevered, and is captain on the world's largest airliner, the Airbus A380.

The average new-hire age at many companies hovers around 30 to 32, but it's dropping. It used to be around 34. Having said that, I've taught weather to several new hires who were in their 50s.

The percentage of female pilots for major airlines is presently about 5 to 6 percent and slowly climbing. The drive to recruit females is huge, with an abundance of scholarships and bursaries available. Sadly, a handful of airlines have yet to hire their first female pilot, but the list is shrinking. Sometimes when flying with female pilots, I watch in awe how they conduct themselves. They make it look so easy. Many think one must be blessed with a "John Wayne" persona (heck, they named an airport after the "Duke" in Orange County, California), but modern-day pilots come in many packages.

The youngest age to have a student pilot permit is 14, but you must be 17 to complete the private pilot license and 18 to get a commercial license. Yes, you can fly an airplane before driving a car! You don't have to have incredible math skills, nor do you need to know calculus or high-powered physics or possess an engineering degree. The basic skill is the ability to read. If you have the gumption, desire, dedication, and right attitude, you will succeed. But that applies to most things in life. As one ornery past boss of mine claimed, "You can teach a monkey to fly, provided you have enough bananas." My wife dislikes me reiterating this because she thinks it demeans the profession.

It's a proven fact, a student pilot taking a little more time to grasp things tends to end up being a better pilot. Society concentrates too much on high marks and overachievers, but well-roundedness in the flight deck wins out. I've also said this about the medical field and life in general. Bedside manner is equivalent to CRM (Crew Resource Management) in the flight deck. I've flown with brilliant people, but sometimes super-intelligence breeds eccentricity, which makes for a fiddly flight deck. Just my observation. Years ago, many pilot interviewers, before the HR department won out, would silently ask this about the candidate: "Could I spend 10 hours in the flight deck with this person?" Think about it, could you spend time with that snoopy annoying neighbor, your mother-in-law (I get along great with mine), or just someone who rubs you the wrong way, in an office the size of a closet for a prolonged period?

Starting salary for an airline pilot is in the $35,000 to $75,000 range, but growth and opportunity is conducive to quick advancement. You are not going to be rich at the beginning, but you will eventually earn a very comfortable wage. I have busted through the $300,000 Canadian cap, but remember, it's taken most of my career to do so. To achieve the minimum pilot qualifications requires about $80,000 to $100,000 in training costs when taking the flying club route. Then you'll need to find your first job to build time, but the journey is worth it!

Some of the hoops for initial hiring at many airlines: a video interview, an in-person interview, a written psych evaluation, a medical, plus a possible cognitive test. References and security checks are done and reviewed before going to a hiring board.

DAPPER AIRLINE DUDS: DO I GET A UNIFORM WHEN I SIGN UP?

For an airline pilot, another perk is the uniform. (A uniform is a must for other departments within the airline, so if you can't decide what to wear to work, then an airline job is for you. Plus, you save a ton of money on buying a wardrobe.) Most airlines pay for your airline attire, but some only pay for half. The clothes are expensive, so some airlines will charge you for the first set of airline duds and pick up the tab thereafter. And yes, a pilot uniform turns heads. But remember, when you don your uniform, people are watching. There is strict protocol set out by the company, and they too are watching.

✈ Under the Virgin Atlantic livery, flight attendants have a unique red theme to their snazzy, eye-catching outfits.

Years ago, in London, England, I asked a Virgin flight attendant (I was asking for a friend) whether their entire uniform is red. "Yes," she brightly said, "right down to the red thongs!" Guess what I envision as the Virgin flight is barreling down the runway? ☺

✈ The topic of whether a hat should remain part of the uniform pops up now and again among pilots. Those with a full head of hair tend to want them gone, but those of us who are follicle-challenged like our hats.

✈ Unwritten code for a pilot having a dent or crease in their hat is "bird strike," as in, "Looks like your hat has a bird strike."

PILOT PERKS.

Everyone working for an airline receives passes to travel at their will, but they are not confirmed. If you have time, patience, and perseverance, you can travel the world cheap without adhering to a strict schedule. My family gets them too. Be forewarned, they can cause contention amongst family, plus you may be making new friends if your company offers "buddy passes." Personal passes are not entirely free because they do attract taxes and airport improvement fees.

Pilots also have jumpseat reciprocal privileges. No, you don't sit in the flight deck (although sometimes you do) but are offered a seat in the back. A pilot could bum a ride on a UPS cargo flight to Paris for the weekend. Then there are professional reciprocal discounts found at car rentals, hotels, restaurants, and some major attractions. Even a major coffee chain (I like Grande Bold) offers an airline discount at many airports. You will find airline types incessantly asking for "airline discounts." It's engrained in us. But sometimes I get a blank stare, as in, "What are you talking about?" Then there are the dental, medical, and pension perks and the ability to buy company stocks. A cardinal rule is don't buy stock in the company you work for and never invest in airline stock. I didn't heed either advice. I've lost thousands, but as of late, I'm looking stock-market savvy.

FROM ABOVE (WRITTEN SOME 15 YEARS AGO).

Check-in is one hour and 15 minutes prior to departure, but I try to show up a little early, as it is certain I will bump into some colleagues at flight planning. Undoubtedly, the number one question heard in the flight planning room is, "Where are you off to?" Answers range from a 15-hour polar Hong Kong flight to a short, super-busy flight to LaGuardia, New York. We print the flight plan along with the many pertinent weather charts. (Note from 2022: everything is included on our company-issued iPads, but many, including me, like to have some portions on paper.) Sometimes a quick call to flight dispatch, located off airport premises, is warranted to check on changing weather conditions, ride reports (flight conditions), and any maintenance issues. We will never meet these people except on rare occasions. Just like the air traffic controllers. We do identify some voices, but most remain faceless.

It's then off to the gate to settle into the flight deck. Walk-arounds, logbook checks, and briefings to the flight attendants are just a few of the many things transpiring as we ready for an on-time departure. Ramp checks, fuel checks, and inputting the flight plan into the onboard computers are all part of the job description. Nowadays, our flight plan is uploaded via datalink to reduce time and mitigate errors.

Pushback commences with everything abiding to standard operating procedures. Taxiing to the active runway entails more checks. Finally, a "cleared for takeoff" is read back to the control tower. Again, everyone sticks to the script as we begin the takeoff roll. Even after many years on the job, there is still a tingle of excitement when the takeoff thrust is set, confirming, "We are going flying!"

One way to describe my job is moments of sheer terror coupled with hours of boredom. As those prolonged hours pass by in the flight deck, it gives time to reflect on the vastness and beauty of this country from above, with so many landmarks and communities below conjuring up numerous "ground" memories.

During the wee morning hours, while setting course for Europe over Newfoundland, the many lighted communities below bring back memories of countless fun-filled layovers and of distant relatives scattered across the "Rock" (Newfoundland). On return flights, daylight gives way to spectacular sights of icebergs meandering southward in the frigid Labrador Current. Sights emblazoned in my boyhood memory when I sailed with my father on coastal supply boats before the Trans-Canada Highway was built. Maybe this was the impetus that gave me the travel bug? I witnessed so many fishing communities, each with its own uniqueness, but unfortunately many memories are now blurred by time.

As I traverse the ragged mountains while descending into Vancouver, either from an Asian flight or from the east, many tree plantations come into view. It makes me thankful I'm no longer bent over, busting my back during arduous spring tree planting, counting every tree to earn money for university and flying lessons. I visited many towns while on the move to new plantations, sometimes with youthful recklessness. I'll never forget the sight of a fellow tree planter surfing the hood of a truck en route to Banff, Alberta, cruising at 60 mph (100 km/h).

Even my flight lessons allowed me to see some neat spots in Canada. Flying lessons started on the east coast in Halifax, Nova Scotia, later finishing in Victoria, B.C., with summertime commercial pilot training in the town of Gimli, Manitoba, which is infested with fish flies. Few believe me when I tell them of the fish-fly stories.

The sight of blizzard-like conditions of awkward fish flies colliding under night lit streetlights always returns to mind as I look below toward Gimli.

Large cities seem so tranquil from above, devoid of noise, commotion, crime, smell, language barriers, cultural walls, and politics. I learned when flying over fireworks launched from cities below, without sound and energy from spectators, that they just aren't the same. Cities can appear pristine with fresh-fallen snow or gloomy under a blanket of self-induced smog. The greens of spring contrast with the dull browns of late fall . . . truly habitats of contradiction. Traffic seems almost organized, though the perpetual beams of headlights emanating from suburbs during early-morning commutes make me appreciate the freedom of flight.

Seasons display their spectacular sights: leaves turning on the hardwood in the fall, snow boundaries etched by recent winter storms, the northern lights dancing in the Arctic nights, spring fog hugging the Atlantic coast, and monstrous summer thunderstorms in the prairies. Weather covers a gamut of extremes, giving me, as an airline pilot and ex-meteorologist, a new respect as to what weather can dish out.

Years ago, invited passengers could capitalize on the vantage point from my first-class front-row seat; however, a very expensive, heavily armored locking door now severs those chances. Even the flight attendants pay fewer visits.

To see Niagara Falls, the CN Tower, the big "O" in Montreal, mosaics of prairie farmland, snow-covered Arctic communities, the Rockies, all from a bird's-eye view gives you an appreciative humility. I've worked on Sable Island as a summer student, partaken in university fraternities in Montreal, encountered the joie de vivre of Le Carnaval de Québec, stressed myself out in flight simulators in Toronto, planted some 400,000 trees in B.C. and Alberta, and toasted the northern lights in Goose Bay, Labrador. Canada is vast, diverse, and loaded with quaint communities and cosmopolitan cities.

Writing this is in the middle of my kid's two-day swim meet in Lunenburg, Nova Scotia. (Talk about hours of idleness for seconds of flurry.) While waiting for the day's races, I met a spry local, Clement Hiltz, 92 years young. He sailed on the original *Bluenose* and served in World War II. I met him in Lunenburg's large downtown gazebo as he readied himself to recount stories of yesteryear to tourists. *Good Morning America* thought the town intriguing enough to air a show from there. To meet someone who sailed the boat found on our dime, and who defended our country abroad . . . now that's Canadian. He spent his lifetime in a shipbuilding town that boasts 250 years of heritage. They painted the town meticulously, and he thinks it's a community second to none in this country.

Next time my flight plan takes me to the east coast, en route to destinations around the world with a large maple leaf inscribed on the aircraft tail, I'll look along the coast to Lunenburg and think of the elderly gentleman I met who had the same feeling of pride — not just for an individual community, but for a country filled with hundreds of them, all with their own uniqueness and distinction.

WALKING THE GANGPLANK.

As I walk the simulator gangplank, I transition from reality to the virtual world. There I'm challenged, observed, and checked. It's a world few people experience, but a necessity for what I do. It is here where I train in the virtual world to ready me for the real one. This world can whisk me from an approach into New York with an engine fire, to a zero-visibility landing in Vancouver — all with the touch of a button. Other nearby "sims" in this cavern of a building put other pilots through their paces: one may be virtually mired in San Francisco fog, the other performing an emergency descent due to a loss of cabin pressurization in Hong Kong, and yet another taking off in a blinding Montreal snowstorm. So many virtual worlds . . . so many scenarios.

After hours of squirming, screwing up as well as nailing procedures, this pivotal box of magical wizardry comes to rest on its powerful hydraulic or electric jacks. The gangplank descends to allow two other pilots their turn. Smiles are exchanged, but they bear resemblance to those seen at funerals. As one captain who flew his last sim said, "It never gets any easier." And I concur! It's a pity, but the toughest part of my job is trying to keep it.

AN EXCERPT FROM CAPTAIN D'S PENDING BOOK — *AUTOLAND*.

(An Airbus 320 is en route to San Francisco, but both pilots have succumbed to food poisoning. It's up to a 17-year-old Microsoft simmer to land the crippled airliner — that is, before two F-18 jet fighters pluck it out of the sky over the badlands of South Dakota.)

"Power loss!" reverberated throughout the flight deck as the crippled airliner labored into the air during rotation on takeoff. Captain Dave instinctively knew what to do as the failed left engine groaned to zero thrust. The asymmetric thrust caused the airplane to veer from the runway, but quick rudder input from Dave's right foot kept it tracking in a straight line, allowing the heavily laden airplane to climb, escaping the catastrophic consequences of plunging into the lurking terrain below. Two engine airliners are certified to climb away on a single engine, but Dave knew one wrong move could easily flip it over on its back.

A crisp "Positive rate!" echoed mere seconds later. Finally, at 400 feet (122 m) above the ground, Dave commanded Bob, his first officer, to engage the autopilot and secure the engine. Bob engaged the number one autopilot, albeit with an unsteady hand, and began the drill, ensuring Dave confirmed his actions. With a two-engine airliner at maximum takeoff weight, now was not the time to accidentally shut down the wrong engine.

"Remember Dave, level off at 2000 feet so we can get the flaps up, we don't need any more drag." Beads of sweat formed on Bob's forehead and upper lip. Sweat also infused his shirt, with the smell of BO permeating the closet-size flight deck. The level off came with a push of a button, with Dave stating crisply, "vertical speed zero." The airspeed slowly increased as the disabled airliner wavered in flight like a bird with one wing tied behind its back.

"There's the speed you need, Dave, to get the flaps up!" As Bob's voice shifted to a higher pitch.

Dave glanced over at Bob, realizing Bob was wound up like an eight-day clock. Dave asked himself why he always got paired with Bob, but then clued in as perspiration formed on Bob's overweight frame — nobody wanted to fly with Bob. No one was willing to jeopardize their license with a fair-to-middling pilot. As one pilot supervisor once told Dave, this company had A+ to A- pilots, but there were a few Bs and Cs out there. Dave realized he was flying with one of the "Cs."

The flaps slowly retracted into the wing's trailing edge. The aerodynamically clean airliner was ready to climb to a safer altitude, allowing Dave and Bob to confer as to their next plan of action. The in-charge had to be briefed, along with a full load of wide-eyed, freaked-out, hysterical passengers. But not before air traffic control was notified of Dave and Bob's intentions.

Dave, the skipper of this European-made Airbus 320, declared an emergency. "Mayday, Mayday, Mayday, AirJet Flight 401 has lost an engine on takeoff, we have 146 souls on board, eight metric tons of fuel, and negative dangerous goods. We want priority vectors back to the airport for an immediate landing!"

"Roger, AirJet 401, the weather in Toronto is down to half a mile visibility in fog. I'll set you up for an ILS on Runway 23."

Bob got to work plugging data into the computers for the instrument-aided landing — a must in these weather conditions. Dave briefed

Bob, and they followed vectors to set up for a straight-in approach. Dave let the autopilot perform its magic.

After what seemed like an eternity, Bob blared out the word "Minimums!"

Dave immediately responded by calling, "Continue!"

The airplane settled onto the runway with thrust reversers promptly selected, enabling the reverse flow from the good engine to aid the airliner to a more efficient stop. The park brake was set, even though the plane was on the active runway, effectively closing the runway, with Dave blurting over the PA, "Remain seated, remain seated!" Dave wanted the crash vehicles — now called rescue vehicles because the word "crash" didn't sit well with the public — to check the airplane over before he taxied off the runway.

That was enough for Harrison Jones to see. A seasoned, deep voice emanated from directly behind Dave and Bob, "That's great guys," as the simulator retreated to its resting position on the pressurized hydraulic jacks. A beeping was heard as the bridge to the simulator lowered. Described by many pilots as the gangplank to misery, it connected the stilted two-story-high hydraulically maneuvered expensive video game to the fixed stairs of the building. The beeping — a familiar sound to all pilots — indicated that recurrent training consisting of two days of torture, required twice a year for all airline pilots, had come to an end.

To leave out any uncertainty, Harrison made it known they had passed. He didn't believe in making them wait to hear the verdict in the debriefing room. "Oh, in case you're wondering, you both passed," Harrison said, grinning, as he opened the simulator door from the virtual world while picking up the notes he'd made during the ride. "Remember to bring all your stuff: headsets, checklists, and iPads. Oh yeah, can you guys do a 'parking checklist' to make sure we didn't forget any switches for the guys to follow?" asked Harrison as he exited the stench-filled high-tech simulator.

Harrison, with six months to retirement, knew the simulator world would still be with him for several years to come. With an ugly divorce and two little ones from his second marriage, retirement wasn't an option. Regulations would clip his wings at 65, but Harrison was in great shape and still had aviation engrained in his blood. Many pilots couldn't wait for their last flight, yet others wanted to continue into their 80s. Harrison was one of those who would go out kicking. He would be able to teach pilots for years to come, but never command an airliner — at least in North America. As one pilot had said years before, "Aviation isn't a passion, it's a disease."

Harrison knew the systems of the Airbus 320 inside out. He was a check pilot supervisor (or check airman, depending what airline you flew for), had done sim training for years, and was well regarded amongst his peers. That carried a lot of weight, as some line pilots thought some checkers were dickheads. Checkers were comparable to army officers at arm's length from the battlefield, with line pilots as the soldiers in the trenches. A line pilot never trusted a checker, especially when they smiled. In fact, one supervisor had a smirk on his face as he ripped up a pilot's license after an unsuccessful simulator session. No wonder he was known as the "smilin' assassin."

Left behind were Dave and Bob to gather the belongings needed during the four hours of cockpit challenges. "This stuff never gets any easier," said Dave, looking at Bob stuffing his flight bag with out-of-sequence, jumbled-up notes and charts. Dave knew it would take Bob quite some time to get things back in order.

"When are you flying next, Bob?" Dave wanted to break the ice because Bob seemed more rattled than usual, and so the question went unanswered. Finally, Dave shook hands with Bob and said, "Good ride." Bob meekly replied with, "Thanks, Dave."

The trio walked back to the Airbus A320 briefing room. The clunking from the hollow floor sent Dave a vivid reminder of the simulator building. The floorboards were easily pried open for the sim techs to access the miles of wires which drove the ten simulators, each costing about $20 million — a high price to pay for safety. Each simulator would be on the go the entire day, giving pilots a gamut of failures and scenarios. As Dave walked by the sim of the Boeing 777 — the largest airplane at AirJet — he wondered what part of the world the crew were in, maybe doing a reject at the Hong Kong airport or contending with a slippery snow-covered runway in Montreal.

Harrison debriefed Dave and Bob in the tight-quartered briefing room supplied with reference books and a cardboard mock-up of the flight deck. The door was closed to drown out the noise of the fans used to cool the simulators as they produced their ever-so-real graphics. The sim was so good at replicating the real world that pilots didn't need to fly the actual airplane to have their license signed. In fact, the first time they saw the real airplane was with a full load of passengers.

Protocol was evident in the room. Though it was unnecessary, Dave sat in the left seat, and Bob the right. Harrison started with Dave.

"Dave, good ride," said Harrison as he signed Dave's license. "One small point — and this is from the checking department. When you declare an emergency, you don't say you lost an engine. An engine didn't fall off. You should say you had an engine failure. I know it's a moot point, but I had to say something. Anyway, the ride was to a 'very high standard'" — the highest the company rated flight tests.

Then Harrison addressed Bob, whose eyes shifted immediately to the floor, "Bob — not a bad ride."

Even though Harrison was overly kind about Bob's overall performance, Harrison knew he could debrief Bob on many facets of the ride, but everything was within limitations. Bob's face lit up. Bob suffered severely from "checkride-itis" — stressing out more than necessary; a tough thing to overcome. It was the equivalent of stage fright for pilots. Harrison knew it and saw the benefit of propping up Bob's ego. Handshakes were given as Dave asked Harrison, "How long will you be doing this thankless job?"

"Until our little one is in college. Looks like I won't be golfing much in retirement," joked Harrison.

The next two pilots walked by the window of the simulator briefing room with their check airman in tow. *There go two more stressed-out guys*, thought Dave.

ALMIGHTY SENIORITY.

A seniority number is everything in most airlines. It dictates the type of aircraft you fly, your domicile, position, vacation allotment, monthly schedule, travel pass priority — and heaven forbid, if there are layoffs, your better number may keep you employed. The week of your new-hire course, even if your wife is expecting, set your butt in the classroom Monday morning if you get the chance. I realize this may be a brazen comment, but I want to stress the importance of this sacred seniority number. I've heard countless stories where pilots delayed coming, with most having major regrets. Go for what is being offered at the time! Don't second-guess the "what ifs." Having said that, some companies are shifting toward status pay, whereby years of service, not aircraft type, dictate pay. But for most major North American carriers, the bigger the airplane and the more senior you are, the fatter the paychecks. Within the major airlines, switching the seniority hierarchy to a more level playing field would be equivalent to telling people to drive on the other side of the road or convincing Americans that Celsius is better than Fahrenheit. It's not going to happen soon.

✈ Acquiring a seniority number is as basic as waiting in line and taking a number. It's first come, first served, and you can't butt in line. At the end of teaching the new-hire class, I offer these words of wisdom: "This company is like a large Ferris wheel, just get on it and enjoy the ride!"

Many believe promotions should be predicated on merit. Nope. If Canadian astronaut Chris Hadfield joined an airline, he would start at the very bottom. Some 23-year-old new hire who truly won the lottery by getting hired ahead of him may be his captain. Even Maverick from the movie *Top Gun*, or a pilot with Captain "Sully" (miracle of the Hudson River) status, or the president's hand-picked pilots flying Air Force One must start at the bottom and look up. If you are seniority number 3000, you don't get promoted until number 2999 makes his or her choice. Talent, qualifications, how nice you are, or how many medals you gained in the military doesn't promote you any faster. This is a shocker to many. To complicate things, there is relative seniority that varies on aircraft type, position held, and base (domicile). For example, a very senior first officer (right seat) decides to go "sideways" to the left seat. They would give up the luxury of holidays, summer vacation, and hand-picking their routes, instead sitting at home waiting for the phone to ring at five a.m. to tell them where they will be heading for the next four to six days. They will start near the bottom, as their seniority number is junior. I was a very senior captain on the Airbus A320 with only about six ahead of me, but when I went captain on the B787, I could no longer avoid working Christmas, nor summer vacation, and probably never will. Seniority is all about timing and luck, and it is sort of like the Snakes and Ladders game.

✈ One aviation adage is to "stay senior on junior equipment" rather than being "junior on senior equipment." (I've done both in my career.) Staying senior allows a better lifestyle. So much so that some pilots will defer promotion until the end of their career or just coast out to retirement extremely senior on a smaller aircraft or in a first-officer position. Yes, they will give up higher pay, but they and their family have grown accustomed to a great lifestyle, or they have other endeavors outside of work.

Nothing, I mean nothing, is more important than this "number." Every pilot knows their "number," yet some pretend they can't remember it. Hogwash! Not buying it. Even when pilots get together, they sniff others out by asking what equipment they are flying, their position, and what date they were hired. Many are aviation detectives. Heck, some pilots walk around with the latest seniority list and have been known to pull it out now and again. The only way to deke around this carved-in-stone ranking is to become a management pilot or a check pilot/airman. You'll see many near the bottom rung of the ladder take these roles, as it gives them more control over their schedule and life. Plus, it's bragging rights. You will rarely see a manager or check pilot/airman toting their overnight bag on Christmas Eve. This Christmas Eve (2019), with 24 years flying for the airline, I flew a red-eye from Vancouver to Toronto, touching down at six a.m. on Christmas Day. I witnessed Santa once again traversing the winter sky.

✈ Every company has their unique way of assigning seniority numbers. Some do it based on age or flight time, while others have agreements with their connector airlines. My company pulls the number out of a hat. Can you imagine drawing number 30 out of 30? But that last position is still ahead of the entire next class. And what about when airlines or smaller companies merge? What a hot potato! I've been there, and it's taken nearly

15 years for the numbness to go away. As a cardinal rule, never talk about religion, politics, or sex during polite conversation; seniority in a merged airline should be fourth on that list.

✈ Recently, seniority number one has left the building at my airline. He sat numero uno for over a decade, but after 46 years of aviating he retired, setting the parking brake for the last time. It's each pilot's prerogative as to when exactly they want to leave, with many hanging on to the bitter end. Mandatory retirement has shifted from 60 to 65, but even this is being challenged, not only because people are living longer and are healthier, but to handle the gargantuan pre-COVID pilot shortage.

✈ A very young junior pilot and a very senior pilot who were flying together told each other their seniority numbers. The new junior first officer looked over to the senior captain and said, "I wish I had your seniority number." But the captain looked back at him and said, "No, I wish I had your seniority." Meaning, if he could only turn back the clock. This brings tears to my eyes, as I am now that senior, crusty, nearly-out-the-door captain.

Seniority numbers are updated every year. Presently, I am number 287 out of 4500 pilots. Every month, everyone moves up several notches due to retirement, but the master list is not updated until the first of the year. It's a cheap thrill to see your new number — sort of like seeing American dollars converted into Canadian pesos.

ANNOUNCEMENTS AND THE CONSTANT DIN AT AIRPORTS.

When I commuted from Halifax to Toronto, the perpetual bombardment of announcements irked me more and more. I tried to get through this annoyance by doing a crossword or Sudoku. Airplane safety announcements rank high in making an airplane ride less enjoyable. On your next flight, notice how many people watch the safety demo or listen to the canned safety demonstrations. If I were president for the day, I would reduce this repetitive banter to about 25 percent — or less. It does very little in the name of safety. If a person can't figure out how to fasten their seat belt, then there are other issues at play. This rule was implemented before seat belts were mandatory in cars. It's unnecessary to reiterate, "to fasten, insert the flat metal fitting into the buckle, adjust to fit snugly, and simply lift the buckle to release." They must think passengers have been hiding out in the deep Amazon jungle and find airline travel a complicated endeavor.

Many airlines are making light of the mandated safety videos and are getting a lot of attention for creativity. Air New Zealand and British Airways come to mind for making safety demos more palatable. And there are numerous flight attendants seen on social media acting out their frustration with the repetitive jibber by ad-libbing. Japan Airlines is now hosting an animated reality safety demo that includes the perils of taking your carry-on during an evacuation, considering a recent evacuation where people perished due to jammed aisles. Plus, they include a pictorial of what happens when seated without your seat belt fastened during severe turbulence. For such a conservative country, their safety video is miles ahead of most.

We tell passengers to "store heavier items on the floor and lighter articles in the overhead bins." Try fitting your oversized 30-pound carry-on bag under the seat. Yes, I realize these archaic rules stem from airline authorities, many written in the post–World War II era.

There is a person in cubicle 34 at the governing authority responsible for cabin announcements, and there is no way they will be changing the script on their watch. Because of it, grab a crossword.

Then we must be politically correct with languages. For an overseas flight, this may mean listening to torturous noise levels three to four times. In Canada, this requires everything to be blared out in both official languages. Look around at the demographics of present-day flights. Many speak another language or just don't understand. In the good ole days, flight attendants would go through the cabin ensuring everyone had their seat belt fastened, their seat in the upright position, and items stowed. Automation sometimes breeds complacency and annoyance. It's like riding on the London tube, where you are reminded to "mind the gap," or visiting California, where everything you eat or do is cancerous. We tolerate it, so it will stay. Although I do appreciate the City of London telling everyone to "look right," painted in large white letters, prior to stepping onto the street. It saved me a couple of times while in a jet lag stupor.

Boarding too can be much quieter, but some gate agents believe they are the airport lounge DJ. And heaven forbid if someone compliments them on their announcements, because that constant blather will reach storytelling length. Then you have TVs blaring in many airports, as if passengers want to hear a continuous bout of CNN. Some airports around the world have it figured out by prevailing in silence. Silence mellows a passenger. Sure, if you have an important announcement or something out of the ordinary has popped up, let your customers know.

✈ Speaking of announcements, she is known as "the voice." Carolyn Hopkins is a woman in her 70s living in Maine, and her voice is heard in over 50 countries and hundreds of airports throughout North America. Many think the "voice" is computer generated. Not so. You can go online and listen to the interviews she has done demonstrating her talent if you are missing the ambience of an airport.

✈ Peter Mansbridge (the Canadian equivalent to Tom Brokaw) apparently was discovered while making boarding announcements in an airport in remote Churchill, Manitoba. He headed CBC's *The National* newscast for years.

SECURITY THEATER — THE MADNESS OF IT ALL.

Post-9/11 security measures have taken the fun out of flying for many. Any irregularity is immediately challenged; airlines and airports have lost their sense of humor. True, security is imperative, but some procedures are ridiculous, pointless, and only there for show. Yes, everyone is a potential threat, but when I see the elderly and the mentally/physically challenged treated with such scrutiny, I ask, "Is this necessary?" Those poor people. Not sure how many over the age of 75 with walkers could overpower an airliner. Sometimes common sense does not prevail in the line of security. But as more and more people travel, common sense must rise, and I do see some glimpses of change. Recently, not everyone has to remove their shoes. Billions of shoes have been removed and put back on, all because of one unstable person. Now and again, I still must remove my shoes and belt just to show the entire waiting room filled with passengers that the captain is safe to fly with. It's like, "Hey look passengers, we are so thorough and fair, we are challenging the skipper of your airplane. Things are safe now!" You'll see me staring at the ceiling or wall, waiting for an employee taught to say "bonjour" with a thick Anglo accent go through their procedure in saving the world. It's like getting a needle and told it's not going to hurt.

> ✈ During a standard routine of pushing a "randomizer button" to see if we will be given exemption for bag examination, both the first officer and I hit "red." That means overnight and flight bags on the belt, laptops out, tunics off, airport ID off just in case they need to take it, pockets emptied, and hands probed by a mystical wand with a swab to determine whether we have been playing with dynamite. (I wonder how many millions of swabs have been used and whether they actually work?) The first officer, being a little grumpy, curtly asked while having his flight bag searched, "What are you looking for exactly?" The security agent retorted, "Anything that would take down an airplane." Enough said.

"Now that the horse has bolted, it's time to close the barn doors" comes to mind. Much of the concentration is on passengers; they are deemed the enemy coming through the front door. But what about the back door? It reminds me of the history lesson on the fall of the Fortress of Louisbourg in Cape Breton, Nova Scotia. Deemed a superior French seaward defense, it had all its guns aimed toward the sea. The British lugged cannons through swamps behind the unarmed land side and overtook the fortress. September 11 didn't happen because of a lack of passenger security. We should have looked at the back door and adjusted our strategy for the element of surprise. A case in point: on September 10, 2001, a day before aviation imploded, I sat watching CNN at three a.m. due to jet lag insomnia on a Frankfurt layover.

CNN aired a documentary on Bin Laden, stating that he was up to something nefarious but they were unsure what exactly.

> ✈ While passing through American customs, flight crews were being sent to "secondary" for random searching. There, the customs officer asked what I had in my flight bag as far as food. I mentioned a lowly granola bar. This already tall, intimidating man suddenly grew a foot taller, with his chest beginning to heave and his reading glasses slipping down his nose as he read me the riot act. How dare I not declare this nutty granola bar! I too retorted by sliding my reading glasses down my nose to read the small print. Luckily for me, another pilot entered with an unclaimed sandwich. I was released with a stiff warning. Funny, the exact same granola bar brand was for sale at a kiosk at my departure gate. The games we play.

BUTTING AHEAD OF THE SECURITY LINE.

At most airports, there are separate lanes or areas for aircrew, but for those airports that don't have this option we usually have permission to pass ahead. After all, if we waited in line, your flight would be pushing back late. Now and again, this raises eyebrows among passengers, especially those waiting in long lineups, because they too wonder why we don't have dedicated lines. It's embarrassing and imposes an uncomfortable feeling amongst aircrew, so you'll find me apologizing now and again. Not sure why airports can't accommodate us separately; however, it's part of the big equation only a few know. The least favorable procedure is found at Heathrow, London. We must disembark from the crew bus, unload our bags, pass through a separate security building (they also tend to be the most thorough), and reload our bags onto the bus, all while the bus is being searched. The one positive note about this procedure is we get driven directly to the gate.

> ✈ While slugging through the Heathrow aircrew security lineup, two flight attendants were pulled to the side and asked to open their bags and expose their lives to a stranger. One flight attendant casually looked over to her co-worker's opened suitcase and commented on the unique long, cylindrical hair-curling iron atop her overnight clothes. Seconds passed, and then shock appeared on her face as she realized it wasn't a curling iron. Ahem.

I'd better stifle my thoughts regarding liquids and gels, or checking laptops. Enough said about security and customs, let's get back to the aura and fun of flight. Here are some hints to make your travel a tad more stress-free . . .

EXPEDITIOUS AND TURBULENCE-FREE TRAVEL.

Keep the suitcase light, and if possible, do carry-on. Most airline employees tote only carry-ons.

You avoid lineups at check-in and avoid the "carousel round-up" on the other end. When my family of five traveled, every one of us made it easily for a week or longer with just carry-on.

Use a roller bag with solid handles and good wheels. Quality counts. It's what flight crew use, and you can easily portage a smaller secondary bag with the aid of a baggage hook. Plus, they balance each other. Heck, it was a pilot who invented this must-have travel companion. Some passengers are now opting for small duffle bags instead, because much of the weight of a carry-on is the suitcase itself. Some passenger bags can be heard throughout the airport with squeaky or "square" wheels. Sometimes all it takes is a little WD-40. Aircrew, now and again, opt for new bearing-free roller wheels.

Flag your bags with a conspicuous tag, ribbon, or personal memento. Have you seen how aircrews individualize their bags? "This is not your bag!" "My bag!" "Do not remove!" "Mine not yours." "Steal my bag, do my laundry." Or they may have Mickey Mouse ears or something else cutely conspicuous. I have my business card in a small pouch. It's as boring as "aircrew" or the overused red safety tag, REMOVE BEFORE FLIGHT. ("Remove before flight" has other connotations when found on PJs or negligee.) This also eliminates the surprise of someone inadvertently taking your bag from the overhead bin. In a rush during my commuting days, I inadvertently took a female flight attendant's bag and I am glad the mix-up lasted only 30 minutes. My bad.

Get a Nexus card if you are a regular at the airport. It's only $50 US for five years, and that $10 annual fee — equivalent to a single Starbucks visit — will kiss those perpetual custom/security lineups goodbye.

✈ I should practice what I preach. Recently my son and I flew to Los Angeles. He had Nexus; I thought I didn't need it. Wrong. He whisked through security and American customs in a speedy five minutes. I took 90 minutes plus, and had they not posted a delay on the flight, my son would have left without me. Nexus, here I come!

Always pack earphones for the IFE (In-flight Entertainment). You will save a few bucks by not paying for a set on board. My wife and kids don't budge without them.

I am shocked by the size of luggage and "stuff" some passengers portage through the airport, sometimes requiring Sherpa-like porters. If you can lighten the load, you can lighten the stress.

For those slightly apprehensive about airline travel, it is advisable to do a trial run to the airport. Yes, most fear-of-flying courses say a rehearsed airport visit is at the top of the list! It means driving to the airport and parking, or finding and riding the correct bus route, or getting on the right subway. Practice makes perfect.

Wear comfortable clothes. Sure, this is an obvious statement, but remember, you may be sitting for quite some time, sleeping, eating, and trekking to the washroom. Plus, it helps with circulation.

Another in the "makes sense" department is giving yourself extra time. Arriving early at the airport reduces the stress. It sure is nice having a relaxing coffee while waiting for boarding rather than having a domestic brawl in public or turning your dream vacation into an Elm Street nightmare. Sadly, I've seen too many time-constrained, stressed passengers.

Some airport terminals can be a mile long, so stilettos don't cut it. Wear footwear that can whisk you ahead of the crowds.

Always, always put your passport in the exact same place so even a jet lag stupor will not hinder you from finding it. This also goes for wallets and cash, but keep them separate from the passport.

Bring reading material, crosswords, or a few Sudokus, in case, heaven forbid, your video screen does not work or your batteries die. It also passes the time prior to boarding or for a freak delay.

Go on the internet and find which terminal you must depart from / arrive at. Perhaps study the layout as if you are in *The Amazing Race* and need to expedite things. Airline magazines include layouts of the bigger airports to help with the mission.

Airports around the world are somewhat standardized. Departures are on the top floor, with arrivals on the bottom. Don't confuse them.

Pack some snacks. I pack granola bars in my flight bag. But be careful, that orange from the USA is now deemed an agricultural taboo if you try bringing it back.

Remember to stow a credit card on your person, and not in the overhead bin, if you want to indulge with Air Canada Bistro, Flight Fuel Menu, or their equivalents.

Be well versed with the airport signage and monitors. It sure adds stress when you are sitting at the wrong gate. If you are heading to France but hear nothing but Japanese in the airport lobby, you are either traveling with a huge Japanese group or at the wrong gate.

Get "checked out" with the "check-in" kiosks. You will be amazed how easy and how fast these ATM look-alikes help. With a Nexus card and carry-on roll-away bag, and by using the self-check-in kiosk, you will show up at the gate before the crew.

Currency kiosks tend to charge a hefty fee. If an exchange is imperative, keep it on the light side. Most ATMs around the world work fine.

Remember to pack your much-needed medicine in a carry-on. Numerous times we've had to deal with those who tucked it away in the airplane's belly. A delay or a weather diversion will make you quickly realize this may have not been the wisest thing to do.

I am surprised how some young families pack their associated paraphernalia. It looks like they brought the daycare with them. I took my three children all over the world without the gargantuan designer stroller. You can too!

Have a pen handy. This helps with custom forms, visitation paperwork, and those crosswords and Sudokus you brought.

Oh, and that two-week dream cruise you are about to take — maybe you should think about flying to the port city a day or two before it embarks. I've witnessed many leave it to the last flight. All it takes is a small hiccup, and you'll quickly find out cruise ships don't usually wait.

> ✈ In a 2019 interview, the CEO of a major U.S. carrier was chastised for saying, "By the time you sit on one of our aircraft . . . you're just pissed at the world." He was giving a factual account of the process of getting to the airport, fretting over parking one's car, the arduous task of getting through security, and the overall stressful environment for passengers. It's not just his airline; it's a fact of modern-day airline travel. It even makes me think twice about traveling, but I'm always glad I ventured out the door.

✈ CHAPTER 2 ✈

BEFORE START CHECKLIST

CAPTAIN (PIC — PILOT IN COMMAND), FIRST OFFICER (F/O) (CO-PILOT), CRUISE PILOT OR RELIEF PILOT, AUGMENT PILOT, CADET PILOT, SECOND OFFICER . . . THAT'S A LOT OF PILOTS!

The captain is the commander, pilot in command, skipper, supreme being, demigod, left seater, grand poohbah with four stripes on their epaulets and tunic sleeves with gold embroidery on their hats. It's their jugular that lawyers, aviation authorities, and company types go after when something goes awry. Media incorrectly denote the captain as the "pilot." The captain occupies the left seat unless they are a check pilot (check airman) or line indoctrination pilot, where they are qualified in both seats.

The first officer is second-in-command, with three stripes. Many still describe the position using the passé term "co-pilot," hence the confusion when the media uses "pilot." The F/O sits in the right seat. Some may think the F/O doesn't have as much experience as the captain. Not so. It's possible that they have more flight hours, more experience, and a higher seniority number than the captain but chose right seat for lifestyle or other personal reasons. But generally, the first officer is junior in seniority and tends to be younger.

The cruise pilot or relief pilot replaces the captain or first officer for crew rest reasons during the cruise phase. They sit in either the right or left seat, but do not land or take off. An augment pilot is usually a qualified first officer and flies when four pilots are required in ultra-long-haul flights. However, some airlines implement two captains and two first officers for the ultra-long flights.

Some airlines designate their cadet pilots (pilots in training), sometimes called second officers, one or two stripes, but not so much in North America. (Years ago, second officers or flight engineers were the third pilots that sat "sideways.") With the shortage of pilots, a cadet program is gaining more and more traction especially in Europe and Asia. Some companies bestow flight attendants with stripes, maybe one or two, and even maintenance personnel are getting striped shoulders. Heck, airport security has them too.

> ✈ As I write this, Canada just introduced its first innovative cadet program, called Jazz Approach. CAE (a huge training company and supplier of flight simulators), Jazz Aviation (Air Canada's connector), and Seneca College (one of Canada's biggest flight schools) have teamed up to provide a pipeline of top-quality first officers. What a deal! And how have times changed. My vintage of pilots would have run down the street naked with our résumés in hand to get our first flying job.

WHO FLIES THE AIRPLANE? IS IT STRICTLY THE CAPTAIN, OR CAN THE FIRST OFFICER (CO-PILOT) GIVE IT A GO?

Many are surprised to hear the first officer is fully qualified to fly the plane from takeoff to landing. They have the same aircraft endorsement as the captain, albeit qualified only in the right seat. Takeoffs and landings are generally split 50/50. Yes, the captain can decide which leg they would like to fly, but I always offer the option to the first officer. It rubbed me the wrong way when the captain started out the pairing flying first without offering. There are a few rare exceptions during low-visibility takeoffs and landings where it is imperative the captain flies. Okay, if the runway is a sheet of ice for landing and the first officer is brand new, the captain may elect to "take 'er in." Remember that jugular reference?

PF (PILOT FLYING) AND PM (PILOT MONITORING).

Now that you know "who is who" in the flight deck, what about duties? This further breaks down into who will be the PF (Pilot Flying) for the leg and who will be the PM (Pilot Monitoring). On the ground, the first officer reads and actions the checklists, works the radios, and starts the engines. The captain initiates the checklists and taxis the airplane. Then, prior to setting takeoff power, the role changes to PM (used to be called PNF — Pilot Not Flying) and PF. When the first officer flies, it will be the captain working the radios and doing the paperwork. The PM also retracts/extends the landing gear, moves the flaps and other controls, under the PF's guidance. Complicated, eh?

DO YOU FLY WITH THE SAME PILOTS?

Pilots flying for the "majors" have a high probability of never flying with the same pilot twice. The major carriers have thousands of pilots, and numerous airplanes and airplane types. A pilot may elect to move to another pilot base, change airplane types, move from first officer to captain, work on special assignment projects, or become ill, so the chances of sitting in the same flight deck are kind of low. Many people do not have this variability with their job; you see the same people day in and day out. Heck, if a first officer takes a disliking to a captain (it happens), then a first officer can "bid around" the captain. It's inevitable some personalities clash. Can you imagine having this option? What a perk! Over the years, I can think of the "rathers" and the "rather nots." I bet you can too!

WHEN DO PILOTS USE CHECKLISTS?

Pilots have checklists for the following: "before start," "after start," "before takeoff," "cruise," "pre-descent," "in-range" (about 10 minutes before landing), "after landing," "parking" (shutdown) and, if the airplane is finished for the day, or if the crew is leaving the flight deck, a "termination" checklist. The number of specific checklists and the nomenclature vary according to the airline, but you can be assured that checklists are performed from start to finish. Checklists are fundamental to the aviation industry — the most regulated industry I know. They virtually eliminate mistakes, oversights, and assumptions. Checklists have made it into the medical profession, with medical staff adapting to this rigor. In fact, some pilots have jobs on the side teaching the way of the aviation world in the operating room.

All aircraft manufacturers supply generic checklists for each aircraft type, but each airline modifies them. An Airbus 320 checklist at Delta Airlines may vary considerably with a British Airways A320 checklist. The basic checklist remains intact, but airlines try to add cohesion amongst various types. An Airbus checklist may inherit some Boeing traits to help stick to the company's SOPs (Standard Operating Procedures). Even the way a pilot flies a Boeing B787 for United Airlines will be different from how All Nippon Airways in Japan does it. This takes some getting used to when flying for a different airline.

There are mechanical checklists mounted in the flight deck, or plasticized checklists, but the trend for newer airliners is electronic checklists displayed on airplane computer screens. If a switch, button, or lever is moved, sensors will automatically update the checklist. Precautionary and emergency checklists help eliminate necessary memory items for the pilot. For example, if a pilot had to deal with a faulty air conditioning unit, the electronic checklist would lead the pilot to troubleshoot and, if unsuccessful, secure the faulty system.

> ✈ Because the flight deck is so regimented, many pilots unknowingly replicate this approach in many other aspects of life. One instructor would ask the class, "How many of you rearrange the dishes in the dishwasher to make it more efficient after your spouse loaded it?" It sounds anal retentive. Most in his class jokingly admitted to this dishwasher regime. I'm also guilty of that exact thing.

WHAT IS CRM (CREW RESOURCE MANAGEMENT)?

Besides checklists making aviation unique and highly regimented, CRM (Crew Resource Management) is yet another step to mitigate mistakes and oversights, and impose leadership. CRM developed from Cockpit Resource Management to include not only flight attendants but also ground crew. It makes people challenge, query, or allow others to suggest things in the name of safety and efficiency. For years, the captain ruled in the flight deck. And because of different backgrounds, religions, beliefs, and a multitude of other reasons, many accidents occurred because the other pilot or crew did not challenge the captain. This has changed dramatically: every airline pilot must take initial CRM training, which is frequently refreshed. Entire books have been written on this topic. It too has migrated to the medical profession. A few years ago, I broke my ankle playing summer hockey. I needed a plate and screws, so under the knife I went. When I was on the operating table, the surgeon performed a CRM ritual with the staff. I was impressed — then my lights went out.

> ✈ CRM is the effective use of all available resources for flight crew personnel to ensure a safe and efficient operation, reducing error, avoiding stress, and increasing efficiency. But as one cynical pilot put it, "It's diplomatically telling the other pilot to go pound sand." CRM hinges on common sense, and as another pilot phrased it, "CRM training is not needed for most. But for those that truly need it, they know enough to play the game and avoid its true benefits."

WHAT IS YOUR SCHEDULE LIKE?

When I tell people pilots generally fly about 80 hours a month, they immediately think, based on the 40-hour work week, that this is only two weeks. They conjure up a two-week vacation every month. But these are actual flight hours, which don't include the prep work before the flight nor the time after the flight. Flight hours for an airline pilot start when the parking brake is released and end when it is set. Monthly schedules are awarded by a computerized bidding system based entirely on seniority. Again, more and more airlines are trying to make it fairer by spreading the flying more evenly. Flight attendants use a similar system. If you want weekends off or layovers in Miami or London, you will get them, provided you are senior enough. Junior pilots are usually dished out reserve (standby) blocks. Meaning they will be sitting home waiting for the phone to ring or living with their cell phone surgically attached. Pilots and flight attendants want to build their hours quickly. They seek productive pairings (schedules). For example, a pilot could fly to Hong Kong and back in three days and rack up over 30 flight hours. You do that twice in the month and you are up to over 60 hours. Seniority is the biggest playing card as far as schedule. A senior pilot can do all their flying in seven to eight days, which is flying 10 hours a day with the junior guys working 18 to 20 days. It doesn't sound fair, but that's the script we abide by.

HOW MANY FLIGHTS DO YOU FLY EACH DAY?

When I flew for the regional connectors, eight legs (flights) would be the max. Most major airline pilots fly anywhere from one to four legs on domestic routes, but the overseas routes — one leg over, rest, and then one leg back — are the norm. For some long-haul pilots, the number of takeoffs and landings could be an issue as far as maintaining competency. Generally, a pilot must land and take off at least three times in three months. That does not sound like much, and it wouldn't be an issue except the conversation comes up prior to flight a lot. If this requirement can't be met, it's into the flight simulator to reset the clocks.

WHEN DO PILOTS SHOW UP FOR WORK?

Generally, a pilot must check in for work about an hour and 15 minutes prior to takeoff, but this varies according to the airline. Some airlines dictate 60 minutes for a domestic flight, whereas pilots will punch in 90 minutes prior for international flights. There is a ton of work to get an airliner airborne, so on a long flight this may be pushing the limits. I give myself a bit of extra time, but a few adhere to their contract to within minutes. "To each his own" comes to mind.

WHAT IS A DEADHEADING CREWMEMBER?

On your next flight, you may be sitting next to a pilot or flight attendant in uniform.

Why aren't they working? They are, in a way, but they are on your flight because they are repositioning to another city either to start work or after they have finished flying. The term stems from a person who attends a performance or sports event or travels on a train or airplane without having paid for a ticket. Sometimes aircrew dress in "civies" so they can imbibe a drink or two if it's at the end of a pairing or their flying does not start until the next day. It's forbidden to do so in uniform. Many contracts stipulate deadhead crew will fly in business class provided there are seats. This does not sit well with some passengers, but it is the contract. Deadheading is sometimes a result of irregular operations due to maintenance, equipment (aircraft) changes, or inclement weather. Aircraft swaps may mean a route flown mainly by an Airbus 320 is now being substituted by a larger aircraft because of high load factors. Pilots are only qualified on one aircraft type. A Boeing pilot can't hop in the seat of an Airbus and go flying, nor is it legal for a turboprop pilot to fly a commuter jet. Thus, with an aircraft swap, a pilot might travel as a deadheader on the flight they were originally scheduled to fly. Deadheading is also used for regular repositioning. Reserve (on-call) crews are frequent deadheaders, as they can be repositioned to start a trip from another crew base. I tend to dislike deadheading, as I prefer to be up in the pointy end of the aircraft.

MY COMMUTING DAYS, THE GOOD OLE DAYS — NOT!

Surprise is the look most passengers have when I sit next to them in full uniform. I can hear them thinking, "Why is a pilot back here in economy class? Isn't he supposed to be flying the airplane? Shouldn't their company give them a nice business class seat?" I used to be a commuter. I chose to live in Halifax, but I flew out of Toronto. There are many pilots who do not live in the city they fly from. Rumor has it some 40 to 50 percent of U.S. airline pilots commute or have commuted.

As I wrote this, yet another full flight closed and pushed back, leaving me at the gate. Luckily there are two more flights that would get me to Toronto in time for work. Living in Halifax but working in Toronto, I'm torn between two cities. Even people committed to an hour's commute must find it hard to fathom driving to an airport, waiting for a flight on a standby-only basis, and sitting in an airplane for more than two hours, only to wait in another airport to go to work.

As Maritimers, my wife, children, and I tried to assimilate in Toronto for several years. I adapted well, but for my wife, the pull to return home only intensified with time.

Toronto is where my job is, but luckily for my family and me, I had the option of commuting. So we packed up the moving truck and scurried back to the east coast, where things are laid-back, with a more tranquil way of looking at life. I'm not alone in my civic indecision; there are many more commuting from various cities. I'm just glad my commute didn't involve another country, as it does for some.

> ✈ It's been said, "The best thing about being an airline pilot is that you can live wherever you want." It has also been said, "The worst thing about being an airline pilot is that you can live wherever you want."

While some of us travel "incognito" (out of uniform), there's a good chance that the next time you fly you'll run into a pilot commuting to work. Commuting is a way of life for many in the airline industry, and not just pilots. Flight attendants, aircraft mechanics, ramp handlers, and ticket agents have all been known to commute.

And don't assume commuters just show up at the gate and grab their assigned business class seat. Commuters fly standby. Sometimes "standby" can mean "standing" there and waving "bye" to the plane as it pushes back. If the flights are full, commuters wait for the next, always with a plan B. Weather is a big issue. I always checked forecasts to see if anything that might keep me from getting to work on time lurked on the horizon. Many of us have a commuter pad, or as one pilot jokingly put it, a crack house — think fraternity-style living at the commuter's expense in their work city. Other commuters realize this will be their way of life to the day they retire, so they opt for better accommodations. As for me, I stayed at hotels near the airport. I know of several pilots who commuted most of their careers. Even though it can be challenging, there are always compelling reasons to keep doing it: better schedules, more desirable airplane types and routes, a better or more economical place to live, or a happier spouse.

As a follow-up, we moved back to Toronto, and I will fly out my career here. For me, having commuted twice in my career, I can unequivocally say life is too short to commute. But many I fly with will be waiting for a flight to start work or return from work.

DO PILOTS GET SPECIAL MEALS?

If you think two pieces of melba toast in cellophane and a sandwich with a layer of butter thicker than the sole slice of ham, accompanied by a plastic container that lost its lid covering dubious-looking couscous, is being fed, then yes, we get to indulge. Sometimes snacks are boarded, but more and more pilot contracts are opting to be paid in lieu of digesting a delectable crew snack. On international flights, we are boarded crew meals, but if we are patient and there are extras, we get to indulge in leftovers from business class. We are told we shouldn't eat the same type of meal and we should eat at different times from one another. Sadly, sometimes the only snack is "pilot pellets" — packaged almonds or mystery nuts. Again, if we are patient and the flight attendant is tolerant of pilots, we get to indulge in upgraded heated nuts and perhaps cheese trays from business class. But the planets must be aligned for such elegant indulgence. We tend to be consumers of coffee and are allocated bottles of water. With some of our hefty physiques, one would think we are fed non-stop, but that's from sitting on our butt for long periods or solving the world's problems in greasy spoons on layovers. You can ask any international pilot where to get the cheapest beer and a meal in most countries, and domestic pilots where the best burritos are in Denver's airport and where cheap gyoza is found at Vancouver's airport. Special meals are boarded for pilots with dietary concerns, and I must admit, they tend to look better than the melba toast option. Shellfish and shrimp are off the menu, but fish meals are okay.

WHY DO SOME AIRLINES OPERATE SO MANY BRANDS OF AIRCRAFT?

Because large airlines fly to numerous countries and cities, aircraft are picked to best serve certain routes. The 100-seat Embraer 190 wouldn't be used for a transatlantic flight to Paris because it doesn't have the necessary fuel capacity, nor would it make money, with few seats. Alternately, the fuel burn on a Boeing 777 would be too great to justify using it on flights between Boston and New York. But you may see the big airplanes on the short routes because it is probably a reposition flight. My airline includes seven different types of aircraft to accommodate different route structures and airport parameters. As well, connector airlines fly even more varied aircraft. While smaller airports can't handle large jets, others limit landing slots, so we fly larger airplanes to balance the demand-to-cost ratio of having permission to land in those coveted spots. I used to fly to Barbados on the Airbus

A320, but the Airbus 330 or larger is used during peak winter months to accommodate greater passenger volume. The ultra-long-haul polar Hong Kong flights are mostly flown with the Boeing B777 or the B787, which carries enough fuel to send a Honda Civic around the equator about 60 to 80 times.

WHAT DO PILOTS LOOK FOR IN THEIR VISUAL INSPECTION?

A "walk-around" must be done before every flight, with nearly 120 items on the list of things to check. During this "kick the tires before we light the fires" procedure, we check for tire wear, evidence of leaks, faulty navigation lights, damage to engine fan blades, etc. Many vehicles approach an aircraft during ground servicing, so we check for dents and bumps and ensure access doors are closed. Some airlines use a bump inspection system called BINGO (Bump Inspected Now a Go) stickers to acknowledge the "love taps." This walk-around technique is taught during initial training for any pilot, and it's customary to walk around the airplane in a clockwise direction. I've done thousands of walk-arounds, with a few performed in the opposite direction just for a different perspective and challenge. You will see pilots wearing a bright fluorescent safety vest during this check, and during winter, we scrutinize the exterior for snow and ice. On overseas flights, walk-arounds are performed by maintenance personnel for more in-depth checks. Can you imagine how much more safe driving would be if we always "kicked the tires" before heading out on the road?

I SAW AN AIRCRAFT ON THE GROUND WITH FLUID GUSHING FROM THE BELLY. SHOULD I HAVE SAID SOMETHING?

That gushing fluid originated from the drain masts. Whenever one uses the galley sink (flushing old coffee) or the lavatory sink, the liquid does not go into a holding tank but goes directly onto the ramp. Akin to what railroads used to do even when you flush the toilets: the contents go directly onto the tracks. For airliners, only gray water flows onto the ramp. Many ramp attendants get upset because they can get "peed" on. During my ground school class for weather, I show a picture of an airplane "peeing" on the ramp. I advise never to walk under the drain mast, to avoid the onset of this unique precipitation. Incidentally, these drain masts are conspicuously painted with red-and-white stripes to ensure no one hits them. And they are electrically heated when operating at the frigid flight levels.

WHAT'S THE SIGNIFICANCE OF THE IPAD? WHATCHA GOT IN YOUR FLIGHT BAG?

Gone are the days of perpetual paper amendments. Hallelujah! Now, with a couple of clicks, amendments are done almost instantaneously. At one time, it was imperative we amend all the charts, including the airports we never flew to. It's no surprise some pilots hired their kids to do this tedious task. These iPads have also replaced many of the manuals in the flight deck, lessening the weight, and they are where we now download our flight plan. But most pilots, including myself, prefer a paper copy of at least the first section of the flight plan. Prior to the iPad, our flight bag was laden with navigation and airport charts. Some bags weighed 30 pounds plus. Many pilots now portage food and snacks, but only if it's not an issue with local customs. Those bulky rigid rectangular leather pilot bags, sometimes plastered with aviation stickers, are diminishing to smaller ones or being replaced by backpacks. I'm deciding whether I even need a second bag, but it does add balance to my roll-away bag.

THE AVIATION ARMY — GETTING AN AIRLINER AIRBORNE.

It takes an army of behind-the-scene teams to get an airliner airborne.

Without much effort, I could rattle off 60 departments, or 80 if I really tried. Think about it: there's pilot training, flight crew schedulers, fueling, commissary (catering), air traffic control, maintenance, flight dispatch, ramp attendants, deicing, etc. I've been at Toronto Pearson when the fuel hydrants froze (no fuel available); stranded in Vancouver when a major blackout hit Ontario, causing a void of flight plans from flight dispatch; and grounded in the Caribbean because of labor unrest with air traffic control. Pilots are the endmost managers of this profusion of teams. I know everyone is doing their best to get you to your destination safely, making my job that much easier. Over the years, I witnessed some pilots uncomfortable with bestowing others with responsibility. They would constantly challenge flight dispatch on the routing and fuel requirements, challenge the placement of cargo, challenge their first officers, challenge the weatherman's prognosis. Don't get me wrong, we are paid to challenge, but if a pilot did it every flight leg, they would never push back on time.

> ✈ The "out" and "in" times are predicated on when the captain releases or sets the parking brake. Times are immediately transmitted via datalink. Pilots can leave the gate ten minutes early; anything sooner requires an okay from flight dispatch. Airlines around the world are obsessed with OTP (On Time Performance) so don't think they take things lightly.

Every department adds to the equation. If you take one away, a potential delay may ensue. As pushback time approaches, the cargo doors close, weight and balance figures arrive via datalink, the in-charge flight attendant confirms cabin readiness, the ramp attendant plugs in, ready to push us from the gate, the "before start checklist" is completed, the jetway retracts — all culminating within scheduled departure time.

When you nestle in your seat for departure, try to imagine the multitude of departments, not just within the airline, but in the entire aviation infrastructure working to get you safely airborne.

Here is a partial list of the army:

1. Pilot training
2. Pilot medical department
3. Pilot crew scheduling
4. Pilot payroll
5. Pilot managers
6. Pilot recruitment
7. Pilot licensing
8. Pilot uniforms
9. Flight attendant training
10. Flight attendant crew scheduling
11. Flight attendant managers
12. Flight attendant payroll
13. Maintenance
14. Maintenance stores
15. Maintenance training
16. Maintenance (heavy overhauls)
17. Paint shops
18. Airplane washing
19. Cabin groomers
20. Passenger movement
21. Commissary
22. Fueling
23. Deicing
24. Runway clearing

25. Fire fighting
26. Air traffic control
27. Ramp control
28. Overseas air traffic control
29. Weather Service
30. Jeppesen (airport and navigation charts)
31. Transport Canada / Federal Aviation Administration (FAA)
32. Flight dispatch
33. SOC (System Operations Control)
34. STOC (Station Operations Control)
35. Airport security
36. Corporate security
37. Passenger security
38. Customs
39. Ramp attendants
40. Equipment mechanics
41. Airport authority
42. Commissioners
43. Airport maintenance
44. Customer service agents
45. Ticket reservation
46. IT departments
47. HR departments
48. Finances
49. Management
50. Water supply for potable water
51. Lavatory service
52. Hotel arrangements
53. Ground taxi / bus arrangements
54. Advertising/marketing
55. Load/weight department
56. Cargo

57. Dangerous goods department
58. Live animal transport
59. Documents librarian (aircraft paperwork and checklists)
60. Porters (wheelchair service)

And then more layers upon layers of departments, such as the distribution operations and availability optimization department. I'm still unsure what goes on there.

The butterfly effect — a phenomenon where a small change in starting conditions can lead to vastly different outcomes — is named for meteorologist Edward Lorenz's discovery that the flapping of a butterfly's wings could affect weather in the future. Being a pilot/meteorologist, I would like to introduce the "airline butterfly effect" to denote how something small can affect On Time Performance. Just look at the number of departments needed to get an airliner airborne, with most departments intertwined. For example, a conveyor belt to the baggage sorting system breaks at a hub airport. This causes a backlog of thousands of bags. I've seen this very thing happen a few times. It seems like a small thing, but its implications are costly. This seemingly minuscule "butterfly" creates a largely different outcome with multiple misconnections on the other end. I could write a book just on situations that causes delays. Pilots coin the phrase "going off the rails" to describe situations that airlines call "rolling delays."

IATA (International Air Transport Association) standardized the reporting of delays for airlines, allocating a numerical system from 0 to 99. Here are a few I selected from the long list: code 6, no gate/stand availability; code 12, late check-in, congestion in check-in area; code 18, baggage processing and sorting; code 45, AOG (Aircraft On Ground) for technical reasons; code 93, aircraft rotation, late arrival of aircraft from another flight or previous sector. The code pilots dislike seeing is code 64 — pushback readiness.

READY, SET, GO!

The "pushback" isn't just the moment when the airplane leaves the gate. It's the culmination of teams working together, akin to a live performance — for my airline it happens more than 700 times a day. Pilots complete their ramp check, receive air traffic control clearance, review emergency drills, confirm that they have enough fuel for their journey, and converse with the ground handlers. Flight attendants secure the cabin and attend to passengers' needs. The captain orchestrates all these responsibilities like a conductor from the flight deck. Only with the cabin and cargo doors closed, the passenger boarding bridge retracted, the "before start" checklist complete, and the weight of the plane calculated, can a pushback request be made. When the ramp attendant gives the "okay," the captain releases the brakes and then it's "show time." The last item on the before start checklist — pushback clearance has been granted and ground crews are out of the way — is to turn on the anti-collision lights. Once these bright red beacons on the belly of the plane and top of the fuselage have been lit, engine start is imminent.

WHAT IS THE LONGEST ROUTE FLOWN BY MY AIRLINE?

It's neck and neck, but the route from Vancouver to Sydney totals 15.5 hours from gate to gate, beating out the flight from Toronto to Hong Kong by 15 minutes. The "as the crow flies" distance is slightly shorter on the Vancouver-Sydney route (6741 nautical miles versus 6787), but such factors as routing and winds affect the duration. The Toronto to Hong Kong flight, for instance, usually heads over the North Pole because it's the shortest distance; otherwise, the flight is easily the longest, clocking in at about 16.5 hours. Both routes require four pilots working shifts of two.

DO JET ENGINES REQUIRE OIL OR OIL CHANGES LIKE CARS?

Yes, jet engines require a certain amount of oil. You may see maintenance topping up the levels before flight. Jet engines don't burn oil, but it is needed to lubricate parts. While my car has about four liters (1 U.S. gallon) of oil, the Airbus A320 requires four times that amount and consumes about half a liter per flight hour. Generally, there are no scheduled oil changes for jet engines. During engine maintenance, however, oil filters and seals are checked and changed if necessary. There are monitors on the flight deck for oil quantity, pressure, and temperature, unlike a car, where an oil light is the only indicator.

HOW MUCH FUEL IS CARRIED? ARE THE TANKS ALWAYS FULL?

Many are shocked to hear we do not fly with full fuel tanks. When you were kids, and about to embark on a road trip, Papa Bear would undoubtedly fill the gas tank. But fuel is weight, and weight is money when flying airplanes. I don't think I ever "filled 'er up." Carrying extra fuel requires adding 20 percent of that extra quantity: an extra 1000 pounds of fuel would need 200 pounds just to carry it. To fly from A to B in a jetliner, the fuel required is the A-to-B fuel plus reserve fuel of 30 minutes, and fuel to go to the alternate destination. But some airlines are exempt from having to carry fuel for the alternate destination under NAIFR (No Alternate Instrument Flight Rules), provided the weather is deemed good, with a few other restrictions added. Airlines employ sophisticated flight planning software to fine-tune the fuel requirement to within 100 kg (220 lbs). Just think about it: a flight from Toronto to Hong Kong requires 75 tonnes (165,347 lbs) for the B787 plus 2.5 tonnes (5511 lbs) for reserve and maybe five tonnes (11,023 lbs) to go to the plan B airport. Translating into a total uplift of 100,000 liters (26,417 U.S. gallons). If you have a 50-liter (13.2 U.S. gallons) gas tank in your car, that's 2000 fill-ups.

HOW ARE RUNWAYS APPOINTED FOR LANDING AND TAKEOFF?

The number one parameter for runway selection is wind direction. A pilot always wants to take off and land into the wind. Some find this confusing, thinking a tailwind pushing the aircraft for takeoff would be more beneficial. Not so: we want air flowing from the front to the back of the wing, producing lift — thus a headwind is preferred. When cruising, we prefer the opposite, i.e., a tailwind to help push us along. There is no loss of lift with a tailwind when at cruising altitude. Runways are oriented and built based on prevailing winds. But sometimes the winds change, thus a crosswind — a wind flowing across the runway — ensues. If a takeoff or landing must be done with a tailwind, most airliners are limited to 10 knots (11 mph or 18 km/h). A runway change for a pilot can be simple if time is on their side or can be a major workload as things change. I've flown to some airports where runway changes can occur three or more times. This means some fast-fingered pecking on our flight computer keyboard. A runway change at short notice is a workout for any airline pilot. Remember, we are operating flying computers, and telling a computer what to do takes time.

DID I SEE DUCT TAPE ON THE OUTSIDE OF THE AIRPLANE? TELL ME A HOME DEPOT HANDY TYPE IS NOT FIXING AIRPLANES!

Sometimes an inspection door or the lavatory access panel or a piece of trim requires tape. Its conspicuous grey color against the aircraft's livery (paint job) may raise an eyebrow, but it's safe! I have received photos from friends and concerned passengers depicting this grey patchwork, and I must admit it looks like a handyperson's fix. But what do we do if we are off station and the fuel access panel wouldn't close 100 percent? Do we accept a huge delay to get a part or even worse cancel the flight? No, qualified maintenance personnel will apply heavy-duty aluminum bonding tape known to us airline types as "speed tape." Because this tape is aviation-approved, multiply your price guess by three or four to determine its cost. Sure, pilots snicker when reading the snag in the logbook regarding "speed tape applied," but we know full well it is a legitimate yet temporary fix until the issue can be thoroughly addressed at a maintenance base. There is even a comical aviation website that mimics the speed tape concept: Speedtape Airlines, hosted by Captain Roger Victor.

I SAW AN AIRPLANE PART THAT DIDN'T MATCH THE AIRCRAFT'S LIVERY. WHAT'S THAT ABOUT?

Now and again, you will see an airplane part with totally different coloring. That means the part was borrowed, possibly from another airline. Maybe from the direct competition. The nose cone, or radome (radar dome), where the airborne weather radar is housed, may have a totally different color and can be very conspicuous, like a different-colored beak on a bird. Airlines, because of their diverse airports and large network, sometimes employ third-party maintenance, and even the competition, for various services. This reciprocal nicety sees other airlines frequently help each other out. Yes, there is a display of comradery even with the competition.

HOW SIMILAR ARE THE FLIGHT DECKS ON THE SMALL AIRBUS FLEET (A319, A320, A321)?

If you were to peek in at the flight deck during boarding, you wouldn't be able to tell the difference between the smaller A319, the mid-sized A320, or the elongated A321. The instrumentation is identical; they're all equipped with the same side stick (joystick) and flight deck layout. While the Fin Number (aircraft ID) is a sure way to distinguish among aircraft, I knew I was in a newer version when there were electric seats and pullout tray tables — items the first generation of that aircraft model didn't have.

REGIONAL JETS AND TURBOPROPS: ARE THEY AS SAFE AS BIG AIRLINERS? WHAT'S THE EXPERIENCE LEVEL OF THE PILOTS?

Although some airplanes don't conform to the airliner image, it does not mean they are unsafe. Many of the regional jets and turboprops are unfairly called lawn darts, puddle jumpers, or even bug smashers, but they are a necessity to cater to a niche market. The RJ (regional jet) flies at the same altitudes and just as fast as its big brothers, but with a 50- to 75-seat configuration to accommodate the smaller demand. A turboprop is a turbine (jet) engine with a propeller, so these can conquer the shorter runways and be more fuel efficient at lower altitudes.

I have over 6000 hours flying turboprop airplanes in some of the worst weather in North America. Turboprops only soar to 25,000 feet (7620 m), so it's something to think about when booking that ticket for longer flights. You may be in the bumps longer at lower altitudes. And remember, that's how airliners started, with propellers. Sure, the turboprops have smaller cabins with fewer flight attendants, or the first officer acting as one, and tend to be louder, but they can come with a high price tag. Many of them are fully loaded with avionics (electronic gizmos) with many Dash-8 Q400s having HUDs (Head-Up Displays). True, most pilots start their careers flying these airplanes to build time and gain experience, but they all meet the required standards. Don't expect a jet airliner if you are hopping on a flight from Cape Cod to Boston Logan or from Toronto to London, Ontario. And they are not going away, as regional jets and turboprops make up a huge component of commercial aviation in North America. These aircraft are usually painted with the same livery as their mainline partners.

BERNOULLI AND NEWTON DUKE IT OUT TO EXPLAIN HOW A WING FLIES.

Forty years ago, I learned the theory of flight with Bernoulli ruling in explaining the mysticism of flight. In my first book, I explained flight with only Bernoulli at the helm. If you look at the profile of a wing (in aviation lingo it's called the camber), you will notice the leading edge is much fatter than the trailing edge. As well, the top of the wing bows upward whereas the bottom of the wing is flat. When air flows over the top of the bulged wing it speeds up, thus creating lower pressure, because Bernoulli states if you speed up a fluid, lower pressure ensues. The relatively high pressure found on the bottom of the wing will push the wing (airplane) upward because pressure moves from high to low — spoken like a true weatherman.

But I received a very forward email from a reader averring Sir Isaac Newton's laws of motion are the "explanation of the day." This adamant emailer claimed he could prove a piece of lumber can fly based on Newton's laws of motion. For me, lift was never convincing solely based on Bernoulli's principle. It didn't explain how lift felt on my hand while holding it out the window of a moving car, nor did it explain why airplanes can fly upside down. This can be explained more simply using Newton's Third Law of Motion — for every force there is an equal and opposite reaction. For an airfoil (wing), the airflow being forced downwards pushes the wing upwards. So, the modern-day take on explaining enigmatic flight includes both Bernoulli and Newton in the equation.

WHY ARE AIRPLANE TIRES FILLED WITH NITROGEN INSTEAD OF AIR?

Air contains moisture and freezes at high altitudes, where average temperatures hover at –57°C (–71°F). Nitrogen freezes well below this temperature and contains little to no moisture. As well, tires can warm up quickly upon landing. Nitrogen handles the heat much better than air and prolongs tire life by preventing oxidation and rust from forming inside the wheel.

HOW ARE TOILETS SERVICED?

It's a topic that few want to talk about, but it's a huge component of flight operations. If a washroom (lavatory) is deemed unserviceable, it may not be an issue; however, charts are consulted to decide whether operations will be hindered. The number of passengers and duration of flight will dictate the number of required serviceable lavatories. Toilets are usually serviced during most ground stops on long-haul flights and less so for short-hop flights. There is an access panel near the rear (no pun intended) of the airplane to allow the holding tanks to be sucked of human sewage. Believe it or not, this job is sought after at many airlines amongst the ramp attendants, because if they get it, that becomes their only duty. They drive from airplane rear to airplane rear with possible extended breaks. Special biohazard suits and masks are worn by these "lavologists." And the sewage must be "dumped" (no pun intended, again) at a designated biohazard site at the airport. Having washrooms go unserviceable on a long-haul flight could mean a diversion. Yes, it's that serious.

✈ I still chuckle when I reiterate this anecdote. Years ago, when movement in aviation was stalled, a "pilot wannabe" had been bestowed the task of servicing aircraft toilets to keep his hand in aviation. During one servicing, the suction hose disconnected and human waste spewed all over him and onto the tarmac. Severe cussing ensued. Heard by the pilot walking around his aircraft doing preflight checks, the pilot suggested to the waste-covered lavologist, "Why don't you just quit?" The flabbergasted rampie quickly retorted, "And give up aviation?!"

✈ Not sure if my editor will include this, but as alluded to, toilets (lavatories) are a huge element of passenger travel. On the B787, we have a camera outside the flight deck that overlooks the forward washroom. (No, there is no camera inside the washroom, but some crafty pilots have been known to take a picture of the toilet on their iPads and lay the picture on the video screen in the flight deck, convincing the flight attendants otherwise.)

We can observe lots from our cameras mounted in some airplanes. The B787 and B777 have cameras. First, many passengers take several seconds figuring out how the door works. Back in the day, it would be comparable to opening a telephone booth door. Just push! Then there are people venturing to the washroom in their bare feet or socks. That liquid on the floor may not be water. Think turbulence and poor aim.

And some people have never experienced an airline washroom before. We have had passengers view it as a hole in the ground, and thus standing on top of the toilet and letting it go. Years ago, on certain flights, a dedicated employee would travel explaining how to use a toilet. But some still see it as a space station toilet and opt to use the floor of the galley. Another thankless job for a flight attendant — biohazards. Recently, I was minutes away from diverting to Moscow because a passenger was extremely constipated and their heart rate dropped. After several bouts of defecation (everywhere), she recovered and we continued our 14-hour flight. I could write several not-so-nice experiences about poop on a flight. All I can say — shit happens.

DO AIRPLANES EVER GET UNIQUE NAMES?

Most airlines do not affix a name to the fuselage. But you'll see a teal-colored Canadian airline naming a few of their fuselages, especially among their newly acquired B787s. KLM, Lufthansa, and British Airways have names appended to their fuselages, but British Airways have stopped this practice. Virgin airlines not only playfully give their fleet names like *Daydream Believer*, *Lucy in the Sky*, or *Pretty Woman*, but they also emblazon some fuselages with taunting phrases such as "mine's bigger than yours." Good ole rivalry.

Recently during taxiing, I noticed a unique name on a JetBlue aircraft and wondered if they name their entire fleet. Sure enough, looks like their fleet of 260 aircraft have a "blue" theme: *Blue Skies*, *Wild Blue Yonder*, *Chicken Cordon Blue*, *May the Force Be with Blue*, etc. This blue me away. (Sorry, I had to.)

HOW DO YOU KNOW HOW MUCH AN AIRCRAFT WEIGHS?

The basic empty weight of aircraft varies, so precise weights are taken about every five years or whenever modifications are done, such as new seating configurations. The cleaned aircraft is rolled onto sensitive scales — think large bathroom scales. Fuel, engine oil, and potable water weights are all factored in, as well as the hangar's latitude and altitude. Finally, passenger and cargo weights are accounted for to determine a takeoff weight to within kilograms or pounds, depending on the airline.

WEIGHT AND BALANCE: THE ALLOWABLE TOLERANCES FOR AN AIRCRAFT'S WEIGHT AND CENTER OF GRAVITY.

Starting with a basic empty aircraft, the weight and balance department calculates the addition of fuel, passengers, crew, cargo, and even commissary in determining each flight's projected weight to within 100 kg (220 lbs). Standard weight calculations are used for baggage, depending on the destination. Passenger weights are based on Transport Canada / FAA generic weights, but sometimes airlines modify these weights based on surveys. If need be, a male/female breakdown is used to fine-tune things. Around 10 to 15 minutes before a flight, preliminary cargo load figures are datalinked to the flight deck, indicating the plane's weight and weight distribution — and whether your dog Fifi has been boarded. More precise "finals" (final calculations) must be received before pushback can begin.

Weighty facts:

- Having shed coats, mittens, boots, and hats, the average North American passenger is deemed to fly 7 kg (15 lbs) lighter in summer than winter.
- Twenty-three extra kg (50 lbs) per player is adjusted for calculating the carriage of football teams.
- A member of the flight deck crew is deemed to weigh 95 kg (210 lbs) whereas a flight attendant weighs 79 kg (175 lbs). This includes their overnight bags, but excludes shopping indulgences like several bottles of cheap wine from Trader Joe's.
- Checked bag, domestic flight: 12 kg (26 lbs).
- Vacation flight: 15 kg (33 lbs).
- Fort Lauderdale checked bag: 17 kg (37 lbs) (think cruise ship attire).
- International flight (except Asia): 17 kg (37 lbs).
- Asian checked bag: 20 kg (44 lbs).

ARE THERE THINGS THE AIRLINE DOES TO CUT WEIGHT, BECAUSE AFTER ALL, WEIGHT IS MONEY?

When you take even a few pounds off an airplane that's flying every day, the fuel savings add up quickly. One strategy is not filling the potable water tanks to full on short to medium-long flights. We consult a table comparing flight hours to what we should statistically carry. It's part of our checks. Years ago, someone thought we could save money by removing the ladder on the Airbus A330, which gave access from the ground to the avionics bay. Airlines are always looking to lighten the load: reduced paper weight of in-flight magazines, lighter seats, lighter entertainment systems, no duty-free on certain flights, reduced packaging, and so on. One large American airline reduces weight by supplying one olive fewer on the dinner plate. Bean counters love playing with numbers, especially when large permutations are involved. And when times get lean, crunching numbers is always in the cards. Pilots use several methods for saving fuel, from adjusting the flap settings for landing, to changing when the gear is extended or the flaps are lowered, to taxiing with one engine shut down. Count me in for reducing the carbon footprint, but I've seen pilots get a tad obsessive when "crunching the numbers." Years ago, I flew for "Air Rinky-Dink" — a cargo company. We were told to "sharpen our pencils" when calculating our load figures. Thankfully, the big airlines have zero tolerance for this. There are other ways to cut down on weight, but that would involve diets and challenging the carry-on limits, and I'm not going there.

ARE PASSENGERS GETTING HEAVIER? YOU BET!

Transport Canada amended their standard passenger weights in 2019. Basically, a male passenger weighs 24 pounds more than he did in 2004, whereas an average female has put on 17 extra pounds since then.

In the U.S., the FAA also amended the chubby factor to an additional 20 pounds. In 1980, an average American passenger weighed 25 pounds less. The weight of carry-on bags has also increased. An average male's weight, with a carry-on, has increased from 185 pounds to 200 in the summer and, in the winter, it shifted from 190 pounds to a whopping 205 pounds. A female now weighs 179 pounds on the scale, up from 145 (including carry-on), to a 184-pound winter weight up from 150 pounds. Many complain the seats are getting smaller and the separation is dwindling. It's true, but it's not always the airline seat's fault — ahem. It's also why seat belt extensions are requested more and more. As well, the FAA is suggesting passengers should be weighed prior to a flight.[1]

WHAT IS DATALINK? COMMUNICATING.

It's a telecommunications system that works through a ground-based network of radios, with satellite communication taking up the slack in remote areas. Datalink times are directly connected to the parking brake. When the captain releases the brake, an "out time" is sent. This is crucial for On Time Performance data. Plus, most aircrew get paid by the minute, so the clock starts ticking. When the wheels leave the runway, an "off time" is transmitted then and "on time" and "in time" collected.

In civil aviation, a datalink system known as CPDLC (Controller Pilot Data Link Communications) is used to send information between aircraft and air traffic controllers. CPDLC is also used to reduce chatter and lessen errors as clearances and instructions are sent in text and uploaded into the aircraft's system. All we do is hit "accept." These systems are widely used for aircraft crossing the Atlantic and Pacific Oceans and remote areas, and it all starts with a chime in the flight deck. There is a lot of chiming going on in modern flight decks.

---

1 https://airinsight.com/the-pending-new-faa-weight-balance-rules/

Much of our flying is reading the message, accepting, complying, then canceling the message. Heck, I just created a new acronym — as if we needed more in the aviation world. It's RACC (Read, Accept, Comply, Cancel), much like taking a multitude of orders from our spouse.

CRACKING THE THREE-LETTER AIRPORT CODE.

When booking a flight, reading your trip's itinerary, or looking at the tags on your checked baggage, you'll notice three-letter codes that identify airports. Sometimes it makes sense: BOS is Boston, MIA is Miami. But how do you get MCO for Orlando? Often, especially in Canada, where every three-letter code begins with a "Y," they are illogical abbreviations. For most of us, it is one of the mysteries of travel. I will try to dispel some of the secrecy and unravel this Da Vinci Code mystery of flight.

So, for Chicago, why not CHI instead of ORD for one of the busiest airports on the planet? History, along with geographical locations, names of airports, and personal tributes — with politicians' names ranked up there — are what these three letters cater to. Years ago, the National Weather Service devised a two-letter identification system (blame it on the weatherman) to keep a handle on weather throughout the United States. When aviation was in its infancy, airlines simply adopted the system. However, expansion meant that towns without weather stations needed codes as well, so IATA (International Air Transport Association) created three-letter identifiers for airports around the world. Canadian weather offices associated with airports used "Y," which made them easy to identify as Canadian. For some airports, it is easy to decipher: YVR is Vancouver, YWG is Winnipeg, and YQB stands for Quebec City. But where did they get YYZ for Canada's busiest airport, Toronto Lester B. Pearson? Pearson, by the way, was a Canadian prime minister. There is still some shade of doubt about its true origin, but Toronto's original airport, located in the town of Malton, had been assigned YZ for its Morse code telegraph identifier.

Incidentally, Chicago's ORD is derived from "Orchard Field," and the airport's moniker is a tribute to pilot Lieutenant Commander Edward O'Hare. Orlando's MCO stemmed from McCoy Airforce base. It's neat to know that FFA is for First Flight Airport in Kitty Hawk, North Carolina.

> ✈ Many like to display their airport tags on their luggage. It's like how some skiers keep their lift tickets attached to their jackets for bragging rights. I suggest people remove them, as they could confuse the ground handlers.
>
> ✈ Always be nice to the check-in agent! An irate passenger checking in for a flight confronted a ticket agent and demanded prompt service. His rudeness and gruff conduct never seemed to faze the agent, who remained calm and cool the entire time. Finally, after he left, another agent who had noticed this tense situation asked the cool ticket agent how she remained so calm. The agent just smiled and said, "He may be going to New York (LGA), but his bags are off to Detroit (DTW)."

Airport codes are "need to know" information, and many websites are now available to help bust the code. Interesting permutations can arise; in the name of research, I identified SEX for the airport in Sembach, Germany; FUK for Fukuoka, Japan; and HEL for Helsinki, Finland. ☺

To make things more confusing, ICAO (International Civil Aviation Organization) implemented a four-letter identifier for each airport.

These codes are used for flight planning, aircraft navigation computers, and weather info. To change from the three-letter IATA code to the four-letter ICAO code in the contiguous United States, one adds a "K." For Canada, add a "C" and for Alaska add a "P" for Pacific. You and your luggage may be off to LHR (London Heathrow), but the pilots will input EGLL into the flight computers. "E" stands for northern Europe, "G" is the region (Great Britain), and "LL" is for London Heathrow. Going to Bermuda for vacation? It's BDA, but the ICAO identifier is TXKF. Another challenge for a pilot, knowing tons of airport codes.

✈ When teaching my weather class to new-hire pilots, I go over the capitals of each Canadian province using the three-letter codes. It's a good geography lesson as well, because some think Ottawa (YOW) — the nation's capital — is the capital of Ontario. It's actually Toronto (YYZ).

THE BUSIEST AIRPORTS.

Below are the world's 12 busiest airports according to the number of passengers who traveled in 2019. I have flown to them all, except Guangzhou Baiyun. I've included the IATA (three-letter) and ICAO (four-letter) airport codes.

1. Hartsfield-Jackson Atlanta International Airport (ATL, KATL) — 110.5 million
2. Beijing Capital International Airport (PEK, ZBAA) — 101 million
3. Los Angeles International Airport (LAX, KLAX) — 88.1 million

4. Dubai International Airport (DXB, OMDB) — 86.4 million
5. Tokyo's Haneda Airport (HND, RJTT) — 85.5 million
6. Chicago's O'Hare International Airport (ORD, KORD) — 84.4 million
7. London Heathrow Airport (LHR, EGLL) — 80.9 million
8. Shanghai Pudong International Airport (PVG, ZSPD) — 76.2 million
9. Paris Charles de Gaulle Airport (CDG, LFPG) — 76.1 million
10. Dallas / Fort Worth International Airport (DFW, KDFW) — 75.1 million
11. Guangzhou Baiyun International Airport (CAN, ZGGG) — 73.4 million
12. Amsterdam Airport Schiphol (AMS, EHAM) — 71.7 million
(Source: Airports Council International)

But the busiest airports can also be classified according to aircraft movements — takeoffs and landings. These statistics pertain more to a pilot's perception of "busy." By that measure, most of the busiest airports are in the USA. Chicago O'Hare inched ahead of Atlanta in early 2019: ORD is up to over 900,000 takeoffs and landings annually. Here are the top 12 according to their three letter codes: ORD, ATL, LAX, DFW, PEK, DEN, CLT, LAS, AMS, PVG, CDG, and LHR. Yes, I've flown in and out of all of them.

CHAPTER 3

AFTER START CHECKLIST

WHO STARTS THE ENGINES AND TAXIS THE AIRPLANE?

This is a matter of company procedures. A few years ago, my airline designated it the captain's duty, whereas now it is the first officer that starts the engines. Generally, the captain taxis the aircraft to the start of the runway, and after landing, taxis all the way to the gate. Most airliners have hand tillers on both sides, but I have flown airplanes for other companies where only the captain had a tiller to steer the airplane on the ground.

WHAT IS A CROSS-BLEED ENGINE START? IF THE ENGINE CAN'T START NORMALLY, SHOULD I BE FLYING?

Engines are started using high-pressure air diverted from the APU (Auxiliary Power Unit). For the B787 I fly, we start the engines electronically and at the same time. However, sometimes the APU goes kaput, so a cross-bleed engine start is required. This entails high-pressure air from a special vehicle called the Ground Power Unit (GPU). Generally, one engine is started at the gate with the external power plugged in. After the first engine is started, the aircraft is pushed back to start the second.

During the entire process, the air is diverted for starting the engines and not directed to the cabin. Translation: it's going to get warm in the cabin.

WHAT DO ALL THE DIFFERENT HINGED SURFACES ON THE AIRLINER'S WINGS DO?

There are lots of moving parts on a wing besides the trailing edge (back of the wing) flaps and leading-edge (front of the wing) slats. Ailerons are hinged rectangular parts on the aft (back) outer portion of the wing that enable the aircraft to bank, or turn; as one aileron goes up on one wing, another simultaneously moves down on the other. Spoilers on top of the wing also play a role in the banking of an airplane, but most importantly, they reduce lift, acting as speed brakes to help slow down the aircraft or quicken the descent. Upon landing, all the spoilers deploy upwards, becoming aerodynamic brakes — think drag-racing parachute.

WHEN AN AIRCRAFT IS ORDERED, HOW ARE ENGINES CHOSEN?

Many are surprised to learn that aircraft engines are not built by the aircraft manufacturer. In fact, the same engine can be found on some Airbus and rival Boeing aircraft. The Boeing 777, for instance, comes with your choice of engine by any of three manufacturers. My airline opted for the world's most powerful: the GE90, by General Electric. Airlines often select the same engines across the fleet to reduce costs and improve maintenance efficiency, but engine performance, fuel efficiency, and routing are also considered when choosing an engine. It's like making a car: General Motors does not make tires, nor does BMW make windshields.

HOW LONG DOES IT TAKE TO GET AIRBORNE?

Every takeoff is unique as far as how long and how fast we go to get airborne. But generally, 25 to 45 seconds would be a good ballpark figure. During my Airbus A340 (four-engine jet) days, the takeoff would be an exhilarating and somewhat anxious 55 seconds barreling down the runway. Many variables contribute to takeoff speeds and times: weight, airport altitude, temperature, runway slope, wind, runway condition (dry runway versus wet, bare versus snow-covered), engine thrust, atmospheric pressure, humidity, etc. For every takeoff in an airliner, WAT data is consulted. WAT (Weight, Altitude, and Temperature) figures are the three major variables. Pilots input certain parameters, and exact speed values are quickly datalinked back to the flight deck. For the B787 I fly, we hit "accept" and everything is uploaded. Back in the day, we calculated our own data. Pilots flying the smaller airplanes with smaller companies crunch their own numbers in the flight deck.

TAXI!

Radio chatter can be fast and furious at busy airports. English is the international language, but sometimes accents, coupled with fatigue induced by long flights, and airport familiarity (or lack thereof) add to the equation. You can visit air traffic control websites to hear the rapid-fire lingo being directed toward pilots that stresses the importance of sticking to the script. Taxiways are denoted by alphanumeric nomenclature using the phonetic alphabet. That's those large letters and numbers you see on signage. Any "hold short" instruction must then be "read back" to the controller in its entirety. Luckily for me, I have a huge screen, a.k.a. the "jumbotron," depicting the runways and taxiways. There is a higher chance of getting lost or taking the wrong turn when on the ground.

At night, you'll see blue lights along the edges denoting taxiways. Runways have white lights every 200 feet (61 m). During low visibility, green lights embedded in the centerline of taxiways help in getting around. The center of taxiways is depicted with a yellow line, and it's a must that pilots stay on the line while maneuvering. Stay on the "yellow brick road." When new captains learn the particulars of taxiing, these words are often heard: If you deviate from the yellow line and hit something, it's your fault. If you are centered on the yellow line and hit something, it's still your fault. In London Heathrow, Dubai, and Seoul, air traffic control instructs pilots to "follow the greens" for navigating on the ground. Pilots love the simplicity of these guiding lights, especially after a long flight.

The runway and taxi rundown:

- Typical width of runways: 150–200 feet (45–61 m).
- Maximum slope of a runway for an airliner: +/– 2°.
- Main surface of a runway: concrete, asphalt or both.
- Runway lights are separated every 200 feet (61 m).
- "D" is Delta in the phonetic alphabet, but not at the busiest airport on the planet. Because Delta Air Lines is based in Hartsfield-Jackson Atlanta, Taxiway D is called "Dixie" to reduce confusion.
- Runways are oriented to magnetic north. The exception to this rule: in the far north, runways are oriented to true north.
- Meaning of the numbers at the beginning of a runway heading: Runway 24 means it is oriented about 240° magnetic north, but it could range from 235° to 244°.
- The reason runway numbers change over the years: magnetic north moves.
- Taxiway signs are yellow with black letters whereas runway signs are red with white lettering.

DO AIRLINERS HAVE KEYS, AND IF SO, DO PILOTS HAVE SPARES?

Keys are not used to start airliner engines; only smaller airplanes, like Cessnas, require a key for start-up. With large passenger aircraft, we go through a procedure whereby the engines are started by switches and levers. Compressed air, supplied by the APU (Auxiliary Power Unit), is used to start the engine turbines spinning. That hissing sound you hear during boarding is the APU located in the aircraft's tail. You may also notice the ventilation system goes quiet as the APU air starts the engines. Compressed air is required for all jet engine airliners except, as mentioned before, the Boeing B787, where the engines are started electrically and at the same time!

PAINT BY NUMBERS.

At one time, most airlines had their own dedicated airplane paint shop. Now, for many, it's more feasible to fly to paint shops around the world. My airline's airplanes are flown to Los Angeles, Houston, Singapore, and Quebec by dedicated NRFO (Non-Revenue Flight Operations) teams. It takes eight to nine days for the narrow-body planes and up to 12 or 13 days for wide bodies to glisten with a new livery. The process requires days of stripping and sanding, about three days to repaint, and two days to apply technical markings, including registration. The renowned maple leaf rondelle is back, with one now affixed on the belly. New aircraft arrive already painted with the dapper new livery.

Painting facts:

- It takes 240 liters (63 U.S. gallons) to paint the large Boeing B777 and about 130 liters (34 U.S. gallons) for the A320.

- Everything is weighed. A new spiffy paint job weighs 550 kg (1213 lbs) on the B777, with a price tag of up to $120,000 Canadian (US$94,000).
- Usually the fleet receives a fresh coat of paint every five to seven years.
- The newer paint is more environmentally friendly and now lasts up to 11 years, compared to standard paint, which lasts six to seven years.
- NRFO (Non-Revenue Flight Operation) teams are fleet-specific. They also fly aircraft around the world to maintenance bases, perform test flights, and ferry aircraft.

PUT ON HOLD. DELAYS — FLOW CONTROL, GROUND STOPS, HOLDING PATTERNS.

Weather is by far the number one reason why airports slow up. ATC (Air Traffic Control) will implement programs such as ground stops or flow controls. If Chicago is in a snow event, departures from Toronto may be instructed to remain on the tarmac (ground stop) or given a specific "wheels up" time (flow control) to be airborne. Airborne flights may be given holding patterns or headings, or asked to change speed to allow controllers to make time. You may see an aircraft sitting in a conspicuous spot on the tarmac waiting its turn to depart to a high-traffic airport. Toronto Pearson may have aircraft parked at the deice facility during the off season, whereas Chicago O'Hare has a "penalty box" for aircraft to wait for a gate to open.

Facts on delays:

- A ground delay program is a traffic management procedure. Aircraft are delayed at their departure airport to manage demand and capacity at their arrival airport.

- Airspace flow programming is a traffic management process in the en route system.
- A ground stop is a procedure requiring aircraft to remain on the ground. It may be airport-specific or related to a geographical area.
- Toronto, LaGuardia, Chicago, San Francisco, and London Heathrow are some of the super-busy airports that implement ground delays.
- Whenever a pilot is instructed to hold in the air, an EFC (Expect Further Clearance) time is included for planning purposes.
- Entering a hold from multiple directions used to be a difficult task for pilots learning to fly. Now, it's as simple as a push of a button.

WINGLETS. AERODYNAMICALLY DESIGNED WING TIP DEVICES TO ENHANCE EFFICIENCY.

An aircraft's wing entails smooth laminar flow, but air at the wing tip moves from the bottom of the wing to the top portion, causing a twisting flow. This swirling vortex induces drag. Adding winglets reduces these vortices, enhancing the wing's overall performance, including fuel efficiency, by about 3 to 7 percent. Another of its many benefits is that it reduces wing tip wake turbulence that may affect other aircraft — akin to how a boat's wake interferes with other boats.

Winglet facts:

- Other names for wing tip devices include wing fences and sharklets (new Airbus term). The new split scimitar blended winglets (dual feather) are found on the B737.

- The entire "wing tip to wing tip" distance on the small airbus fleet (A319, A320, A321) is exactly 111 feet, 10 inches (4.1 m).

- If you're looking at the back of an airplane, the wing tip vortices will spin counter-clockwise from the right-wing tip and clockwise from the left-wing tip.

- The B777 and B787 use a "raked" wing that produces the same effect as winglets.

- The older Boeing 767s have been retrofitted with large 11-foot (3.3 m) winglets. This nouveau winglet is the largest piece of structure retrofitted to a stock commercial aircraft.

- Sometimes you will see "short sticks" attached to the winglets. These are static electricity dischargers.

- Wing tip vortices move downward and outward. Both air traffic control and pilots will ensure a safe distance from other aircraft to keep it smooth.

- Winglets can improve fuel efficiency, increase range, increase payload (more cargo and passengers), reduce wing tip turbulence, give better takeoff performance, and reduce carbon footprint.

WHAT TEMPERATURE IS IT IN THE BAGGAGE HOLD?

Conditioned air is directed from the cabin, so the air tends to be a little cooler by the time it reaches the cargo areas, which are also less insulated than the cabin. Cargo temperatures: the Boeing 767 maintains its baggage hold above 7°C (45°F), but the bulk area (where animals are carried) can be heated above 18°C (64°F). Controlled temperature cargo bins are also available when temperature-sensitive goods are being shipped.

TREASURED FOUR-LEGGED PASSENGERS IN CARGO.

Cargo bays are pressurized with conditioned air, but generally at slightly cooler temperatures as the air flows from the cabin to the cargo bay. Smaller pets in approved carriers have the luxury of riding with their owners but must be registered prior to the flight. The code in the airline cargo world for "live" is AVI, which stands for "animaux vivants." Pets are placed in the bulk cargo hold, not with the containers. Preliminary load data gives pilots a "heads-up" if "live" has been boarded in the cargo hold.

THE CAT'S MEOW (LETTER FROM AN APPRECIATIVE PET OWNER)

Captain Doug,

I wanted to let you know we all made it safely to Seattle. The cats were well behaved, and the crew was awesome (as always); they kept checking in on the cats to ask us how they were. We learned a lot of good tips in the process. I wanted to share them here in case anyone else asks you or you end up doing another column.

Tips: Soft carrier (we got one that expands on the sides; great for more room between flights — a mini apartment) plus pockets. Let the cats use the carrier as a little cave three to four weeks before (ours slept in it all day every day next to me when I worked at home). Take drives with the cats. It breaks the thought of *carrier equals a vet*. You have to carry your cat through the airport scanner, so think ahead about your carry-on (liquids, laptop, and cats) and how to get through security efficiently. Three hundred and fifty milliliters of sand is allowed per cat in a Ziplock.

We got a small litterbox that folds up into a tiny square. Treats, treats, treats. During training, reward the cat every time they slept in the carrier at home. On the flight, reward. Avoid food and water before flight and have pee pads to be safe. Spray carrier with calming spray. Get it from the vet, not online. Ensure you have rabies and certificate of health. We were not asked for it, but it is required and varies state to state.

Greg Adams
Seattle

DO YOU ADD ANY PERSONAL TOUCHES TO THE FLIGHT DECK?

My only eccentricity that I am aware of: I find myself wearing a baseball cap more often, as my aging skin is showing telltale signs on my forehead. Melanoma is a concern for pilots. But I've seen some interesting gadgets adding uniqueness to the flight deck. Some pilots bring little paint brushes to clean up the crumbs and dust. Some airlines supply such options. Many pilots bring their own headset, often a noise-canceling device. But on long-haul flights, most of us remove the headsets and use the overhead speakers.

One unique captain (now retired) purchased a small battery-operated handheld vacuum from Japan to suck up the flight deck dust. He claimed he had the DNA of most company pilots. Years ago, another eccentric captain, known as "Captain Gadget," had an entire bag of tricks. He would strategically place a mirror on the overhead panel so he didn't have to turn his head to talk to the flight attendants. He would color and highlight the flight plan with numerous lines drawn with a large ruler and a handful of colored markers. When I was a new hire, he had me carry this extra bag of supplies.

Prior to one flight, my first officer placed a large, conspicuous red button on the center console with "bullshit" displayed in a big font. If pressed, the button would spew out the profanity as well. Not sure what management would have thought, but he deemed it funny.

I thought about getting a Hawaiian hula hoop dancer doll like you see on the dashboards of cars. Some pilots correlate its hip-swaying movement to the severity of bumps. But I have yet to purchase such a complicated device to depict such atmospheric perturbations.

✈ CHAPTER 4 ✈

BEFORE TAKEOFF CHECKLIST

WHICH IS MORE CHALLENGING, TAKEOFF OR LANDING?

I get asked this a lot. During takeoff, the biggest concern is mechanical issues, whereas for landing it is weather. On every takeoff roll around the planet, airliners reference a decision speed called V1. Think of it as a yellow light in the decision-making process — do we stop or do we continue? Thankfully, my track record for a "green light" has been 100 percent for every takeoff. But that doesn't mean my next takeoff will be as uneventful. We practice these decision procedures in the simulator. We learn how to reject (abort) on the runway, or if we pass that magical V1 speed, we learn to handle the "what ifs" — more specifically an engine failure during "rotation," deemed the riskiest when we enter the flight regime. Pilots practice these V1 cuts in the simulator frequently. For landing, weather, low visibility, slippery runways, short runways, crosswinds, and low-level wind shear are at the top of the list of inherent challenges.

HOW DOES A JET ENGINE WORK?

Newton's Third Law — for every action, there is an equal and opposite reaction — is easily demonstrated by a released balloon propelled forward by escaping air. The jet's large fan blades pull in air, which is then compressed by a network of fans attached to a shaft. Combustion occurs when fuel is added to the compressed air and ignited. Air and hot gases are expelled out the back to propel the airplane forward. But they all operate on the same principle: suck, squeeze, bang, and blow!

A typical jet engine can suck the air out of a four-bedroom house in less than a second.

WHAT IS A NOISE ABATEMENT PROCEDURE?

Many airports are located near residential and other noise-sensitive areas, so pilots must adhere to procedures to lessen the noise footprint. When departing from Toronto Pearson International Airport, we can't make unauthorized turns until the aircraft has reached 3600 feet above sea level (3000 feet above ground). This standard departure path reduces noise near the airport and adjoining areas. At many airports, when runway conditions are met, we don't use maximum reverse on landing because of the loud rumbling. Some airports prioritize runways to lessen noise during quiet hours, while others don't have set noise abatement procedures. Noise footprints of modern airliners are significantly quieter than aircraft of yore. Gone are the days when nearby house windows would rattle from departing aircraft.

✈ I am told some people living near the Toronto airport make it their duty to ensure all flights adhere to noise abatement procedures. I hear there is one gentleman who has a square taped on his living room window. If an aircraft is outside this square, hence outside the noise abatement window, he will call the appropriate authorities.

DO AIRLINERS USE FULL POWER FOR TAKEOFF?

Many are surprised to hear we don't usually use full takeoff thrust. If you want a figure, about 90 to 95 percent are reduced-thrust takeoffs. Modern engines are so efficient and powerful, they can produce thrust beyond what's necessary to get airborne and climb. The procedure is to input an ambient temperature far greater than the actual outside temperature into the flight management computers. Airbus calls it a "flex temperature," whereas Boeing defines it as an *assumed* or "derate temperature." This programs the engines to perform as if the air is less dense because of the faux higher temperature, thus producing less thrust. I've used takeoff power settings at 80 percent of full power. This reduces fuel consumption, noise, and wear and tear on the engines — meaning less maintenance.

WHAT IS THAT BUZZING SOUND FROM THE ENGINES ON THE SMALL AIRBUS FLEET?

After all these years, this distinctive sound is music to my ears. The engines are high-bypass large-fan types with fan blades spinning at thousands of rotations per minute. The "buzzsaw" noise is from the airflow around the fan blade tips transitioning to supersonic. These same engines can be found on some Boeing products, and the sound is different depending on where you sit. Abeam the engines it is most noticeable.

ON TAKEOFF AND LANDING, WHEN DO PILOTS ENGAGE THE AUTOPILOT?

All takeoffs are done manually. There are no auto-takeoffs. The autopilot can be engaged 100 feet (30 m) above the ground in an Airbus and 200 feet (61 m) above ground in the Boeing, but generally we wait a minute or so thereafter. But many busy airports have very detailed departure criteria, such as altitude, heading, and speed constraints, so I find it's best to get the autopilot on to lessen the workload. Plus, there is less chance of busting a constraint. Many of my first officers will ask, "Do you mind if I hand fly it for a while?" I agree to it, but it does up the workload a tad. Recently, one "I'm going to show you how great a pilot I am" flew it from takeoff to cruising altitude. I haven't seen that done in years.

It's a matter of preference as to when we disengage the autopilot for landing. I like to hand fly the airplane starting from 1000 to 500 feet (305 to 152 m) above ground to get the feel of things. However, for autolands, the autopilot performs the smooth landing.

✈ During pre-9/11 days, our company allowed visitations to the flight deck. During a brief visit, a teenager said directly to the crew, "You guys have the easiest job, all you do is engage the autopilot. Flying an airliner looks so easy." The captain quickly retorted, "Can you play the violin?" "No," said the somewhat cocky visitor. "Why not?" asked the captain. "It only has four strings!"

MUSIC TO A PILOT'S EAR.

Music to a pilot's ear is the harmonic sound of power resonating during takeoff, the repetitive bumping sound when the nose wheel overruns the embedded runway centerline lights, the whirring of hydraulic jacks extending or retracting the flaps on the trailing edge of the wing, the hissing sound from the APU (Auxiliary Power Unit) in the tail supplying conditioned air and ground power. And of course, the staccato of seat belts unclicking once we have arrived at our destination. It's all part of the aviation orchestra.

➤ But what is that barking-dog sound? During single-engine taxi, the Airbus fleet uses a hydraulic PTU (Power Transfer Unit) to ensure adequate pressure. The woof, woof, woof barking sound appears to be emanating from a disgruntled dog below the floorboards. Many passengers pick up on this, but I assure you it is normal.

Chimes and more chimes: the dings are from aircrew calling different cabin stations. Two or three chimes indicate landing is about 10 minutes away, and the infamous Airbus chime is when the "no smoking" sign activates after the landing gear retracts post-takeoff. Boeing also has a cabin alert chime when landing is imminent.

You may see hinged panels rise from the wing, called spoilers. These speed brakes are used to expedite descent or to slow down; thus you may hear or sometimes feel a rumbling. A clunk may be heard as the landing gear tucks into the belly or wing after takeoff, and again as the gear extends for landing. Loud rumbling may be heard immediately after landing as reverse thrust is used to slow the aircraft. But when conditions allow, only wheel brakes are used to slow the airplane — hence a quiet touchdown. It is cheaper to wear out the brakes than to use maximum reverse. In fact, many airlines lease their brakes. The price tag for brakes is about CDN$50,000 (US$38,500) on an Airbus 320.

During boarding, or deplaning after the engines have shut down, you may hear loud rattling or heavy clanking emanating from the engines. When a stiff wind is blowing, enough to blow off a pilot's hat when they do a walk-around, the engines gyrate in the wind. Each blade of these large intake fans shifts a tiny bit to absorb and dampen vibrations, preventing stress on the blades. When each blade passes through the 12 o'clock and six o'clock position, it shifts a tad due to gravity, causing a rattle. The stronger the wind, the louder the rattle.

Then there is the gushing of flushing toilets, the clunking of closing cabin doors, the splashing of deice fluid upon the windows, the whirring of cargo doors closing and opening, and the sound of silence when looking upon the Earth from high above. Pilots depend on instrumentation, monitors, and sensors, but sound is also a vital parameter.

THIRTY SECONDS BARRELING DOWN THE RUNWAY IN A DREAMLINER.

Now that we are cleared for takeoff, I position the B787 onto the centerline of the runway, smoothly advancing the thrust levers to takeoff thrust. In some airplanes, thrust levers are called throttles, and like an accelerator in a car, the thrust levers control the power to the engines.

My left hand leaves the tiller that is forward and to the left of my seat, and I grasp the control column. The tiller steers the plane on the ground by pivoting the nose wheel. The first officer calls out a "thrust ref" (reference) as the two engines spool up past 40 percent engine power, forcing everyone back into their seats. After thousands of flight hours, the sound and feel of thrust is still exhilarating; the B787 engines sound like the screeching, charging raptors in the movie *Jurassic Park*. As the aircraft itself fine-tunes the required takeoff power, my right hand stays on the thrust levers in the event we must abort. Protocol states that only the captain can perform a rejected takeoff. In that event, they would expeditiously move the levers to the maximum reverse position. My heart rate increases a few beats and my concentration is at its highest. I am focusing down the runway, ensuring the airplane is on the centerline, using foot pedals to move the rudder located on the tail to keep us tracking on that line. You may hear and feel repetitive bumps as the nose wheel tires track over the embedded runway centerline lights. Meanwhile, the first officer scrutinizes the engine instruments. As the speed and momentum build, this airliner requires only small rudder inputs to keep it straight as it rolls down the runway. Our ears and sense of feel are in tune for anything irregular, but all we hear and feel is a symphony of nearly 150,000 pounds (667,500 Newtons) of in-sync engine thrust.

We are launching from Runway 23 at Toronto's Pearson International Airport, which means our runway is oriented about 230° to magnetic north. However, there is a stiff wind from our right, forcing me to work a little harder to keep the plane straight, but it's well within the limits of this aircraft. The first officer calls "80," indicating the airspeed is 80 knots, to which I respond, "Roger." This mandatory call serves as an airspeed cross-check — ensuring our computer-driven instruments are up and running — and a pilot-incapacitation check. (Airbus uses 100 knots.) While I keep the aircraft on the runway centerline, the first officer concentrates on scanning the high-tech glass instruments, paying attention to the building airspeed. The airplane's synthesized voice curtly announces "V1" — a pre-calculated speed based on weight, wind, temperature, runway, and other factors. In other words, it is decision time: Do we stop? Do we continue? If everything is normal, I remove my right hand from the thrust levers, which means we're continuing the takeoff. A crisp "Rotate!" is heard from my flying partner, based on another pre-calculated speed. I slowly, silently count one, two, three, applying gradual backward pressure on the control column to achieve flight. At first, it seems nothing is happening, but the near-198-foot (60-meter) wingspan has already begun to lift. (An interesting note: the distance of the Wright brothers' first flight was shorter than this plane's wingspan.) As the wings bite into the air, the aircraft's nose smoothly lifts and a small clunk is felt as the main wheels leave the runway. My focus transitions to the HUD (Head-Up Display), which depicts a ton of information such as attitude, speed, altitude, rate of climb, heading, and a "guidance cue" telling me where exactly to point the nose.

This runway, the longest at Pearson, seems short at this point. Slightly more gradual backward pressure is applied to the control column, and the first officer calls "positive rate," indicating the airplane is climbing and it is time to retract the landing gear. You don't want tons of landing gear hanging out in the wind, so I command the gear to go up by saying, "Gear up!" (Think about the resistance you experience when sticking your hand out of a fast-moving car window.) Another light thump is felt as the gear tucks into the belly of the aircraft. Today's takeoff took an exhilarating 30 seconds. This will be the last time I move the thrust levers until seconds before touchdown. The advanced auto-throttle system calculates and sets its own power settings, much like cruise control in a car. The autopilot is engaged and it's time to unwind. It's hard to imagine the landing is 14 hours away.

WHEN MULTIPLE AIRPLANES ARE WAITING FOR TAKEOFF AT THE SAME AIRPORT, HOW IS A PRIORITY LIST CREATED?

Priority is often given on a first-come, first-served basis, so the first pilot to call for taxi instructions usually takes the lead. Some flights must meet more precise departure times, in which case they may be moved to the front of the line. Other flights may be destined to airports with ground delays, so they must taxi to a holding spot and wait. Air traffic control keeps the flow going by factoring in such parameters as weather conditions and deicing requirements; even aircraft size is considered when planning departure sequencing. A lighter aircraft will have to wait a little longer if a heavier aircraft departed first due to wake turbulence. A lot of juggling goes on, and that includes dealing with arriving aircraft.

LASERS AND FLIGHT DECKS.

With modern technology come unique situations. Lasers are readily available, and if shone into the flight deck, it may cause temporary or permanent blindness. Most people do this not knowing its extreme implications, but it is not taken lightly; the perpetrator usually faces heavy fines or even a jail sentence.

WHAT'S THE SCOOP ON DRONES AND AIRPORTS?

Drones are here to stay, but laws and technology must keep them away from airports. There have been several cases where drones have closed airports. In December 2018, hundreds of flights were canceled after a drone sighting at Gatwick Airport, near London. Drones are the modern equivalent to bird hazards. A decade ago, neither lasers nor drones were in the aviation safety dictionary, but they are today's reality.

WHAT ARE YOUR FAVORITE CITIES AND AIRPORTS TO FLY IN AND OUT OF, AND WHY?

I've flown into the world's busiest airports. Vancouver is at the top of my list of Canadian airports for its stunning scenery. At Japanese airports, I admire the ramp attendants lining up in a row waving goodbye until we taxi. My favorite layovers are San Francisco and London, but imbibing a drink in Paris, Tel Aviv, Frankfurt or Copenhagen never gets tiring either. At night, Las Vegas takes the cake because of its brightly lit downtown district. But it's difficult to choose just one favorite; to see the Eiffel Tower or the Windsor Castle from an airplane, or to land on the man-made island in Osaka, Japan, really makes me appreciate my bird's-eye view.

BOEING VERSUS AIRBUS — THE DUAL AVIATOR IN ME.

There are only two major aircraft builders, Boeing and Airbus. Sure, there is Canadian-made Bombardier and Brazilian-made Embraer, but they pale in comparison to the billions of dollars made by the two heavy hitters. Bombardier's very promising C-Series has recently been gobbled up by Airbus and rebranded as the Airbus A220. As I write, this new kid on the block has moved into the neighborhood.

✈ The rivalry between these two goliaths is equivalent to Pepsi and Coke, Apple versus PC, General Motors versus Ford, or squash and racquetball. If there were a room full of Airbus and Boeing pilots, they would repel to opposite sides like oil and water, each thinking they fly the better airplane. I've flown both, and each has its pluses and minuses. Airbus is flown with a sidestick (joystick), whereas Boeing kept the control column fixed between the pilot's legs. I recently met an Airbus crew in Paris and mentioned I'd switched over to Boeing. The first question out of their mouths: "How do you eat?" They were referencing the tray table that folds out on the Airbus. We Boeing pilots are compelled to set the tray on our lap. I could write several chapters on the pros and cons of each airplane, but as the saying goes, "Love the airplane you are with." Truth be told, in a room void of Boeing fans I confess I'd miss the refinement of "Jacques from Airbus."

FOUR ENGINES VERSUS TWO.

All North American airlines have only two engine jets. The queen of the fleet, the four-engine Boeing 747 jumbo, has been retired in North America. No one operates the four-engine Airbus A380 in North America either; that too is no longer being produced. Yes, you will still see them around the world, with flights into the United States and Canada, but they are only visiting. Two engines have won out over four because of efficiency. Both Boeing and Airbus have opted for the two-engine wide-body airliner. Most signs on North America highways depict a four-engine airplane symbol for international airports. Time to modify the signs!

WHAT'S A STERILE COCKPIT (FLIGHT DECK)?

No, this has nothing to do with using alcohol swabs to sterilize the flight deck buttons/switches/controls, although many pilots go through this ritual during flight deck preparation. Over the years, there have been numerous incidents/accidents when the crew diverted from the script at hand and became occupied with items totally unrelated to flying. To quell these sometimes dangerous oversights, most airlines around the world adopted a policy of "10,000 up and 10,000 down," whereby pilots converse only about flying the airplane unless they're at 10,000 feet (3048 m) or higher. Even flight attendants know this is a bad time to call the flight deck, as it may cause distraction. Can you imagine how much safer driving a car would be if we eliminated distractions?

Many airlines will turn off the seat belt sign, if conditions are smooth on the ascent through 10,000 feet, and that's when the flight attendants leap into action. An in-range check, getting ready to land, is when the seat belt sign illuminates descending through 10,000 feet. This equates to about 10 minutes before touchdown. Ten thousand feet is when pilots remove their seat belt shoulder straps and when to ensure they have them on for the approach.

→ **CHAPTER 5** →

CRUISE CHECKS

ROOM WITH A VIEW.

Passenger windows are comprised of two acrylic panes, with the outer pane built to withstand significant differences between the cabin and outside pressure. You may notice a small hole on the bottom center of the inner pane, used for venting. There is also a third plastic pane fitted with a sliding window blind. Flight deck windows are composed of three sandwiched panes of either glass or acrylic and are electrically heated. Recent technology allows aircraft windows, such as the B787 Dreamliner's, to be much larger, and you won't find the small hole on the second pane. Instead, an electrified gel is sandwiched between the two panes. It can dim the windows, making window shades a thing of the past. The B787 Dreamliner has the largest passenger window of any airliner, measuring 47 centimeters (19 inches) tall, 30 to 60 percent bigger than most others. The Concorde had very small windows to handle the huge pressurization differential at 60,000 feet (18,288 m). Most airliners have flight deck windows that open; however, the B787 has a hatch for plan B.

> ✈ If your travels take you to Charles De Gaulle, Paris, there you will see the iconic supersonic Concorde propped up on pedestals in the middle of the airport. Frequently our crew bus takes us right by the display, and I'm shocked how small the human-hand-sized (about four-by-six-inch) windows are. London Heathrow also has a Concorde on display, but it is sitting off to the side, lacking TLC. Pity.

JUST THE FACTS.

How far away is the horizon you see from an airplane? Distance seen from your "room with a view" can be calculated by taking the square root of your altitude (in feet) and multiplying by 1.23 for a value in nautical miles (see the table below). The formula for distance seen in kilometers is 3.57 times the square route of the altitude in meters. You can get your altitude from the onboard moving map display. To find the square root on an iPhone, tilt the phone to landscape mode to display the advanced scientific calculator. I just discovered this. Who knew? The line-of-sight formula also applies to radio reception, radar surveillance, and weather radar coverage due to the curvature of the Earth.

Altitude versus distance seen to the horizon:

Altitude (Feet)	Distance seen (Nautical Miles)	Distance seen (Kilometers)
10,000'	123 NM	228 km
20,000'	174 NM	322 km
30,000'	213 NM	394 km
40,000'	246 NM	456 km

NIGHT FLIGHT.

During a daytime flight, your airplane window affords many interesting bird's-eye views, but if you know where to look, night flights yield even more spectacular sights. Along with the stars and constellations, the dark offers its own amazing phenomena.

The planet Venus is the brightest object in the sky — aside from the moon. Called both the morning star and evening star, it can take on many guises. Atmospheric effects from Earth cause it to flicker, changing its color from green to red to orange to blue. It can look like a bright light that is following you. It's not surprising it's been frequently mistaken for a UFO. While flying east over Russia during the wee hours of the night, we spotted lights ahead fluctuating between green and red. At first, we thought it was an airplane, but it began to climb higher above the horizon, confirming that it was in fact Venus staging a dazzling light show.

Shooting stars are an occasional treat we see illuminating the night sky. Once, while we flew from Toronto to Montreal, a very bright light streaked past the airplane's windshield. Reports the next day verified that a meteorite had landed in New York State, only a few hundred miles away.

If our routing takes us north, we sometimes see another of Mother Nature's spectacular shows, the aurora borealis (or northern lights). When these dancing lights are at their most breathtaking, we let our passengers know — if it's not too late at night, that is. Another phenomenon worth seeing is daylight giving way to night. This sharp black edge, called the terminator, can be seen during eastbound flights.

The moon can also put on great shows, changing size and color depending on the clouds and atmospheric conditions. A halo effect around the moon, known as "holding water or grease around the moon," is caused by high clouds laden with ice crystals — it's a precursor to advancing weather.

On a clear night, great vantage points allow numerous cities to be seen in one glimpse. As mentioned, my favorite city for landing at night is Las Vegas. The multicolored neon lights of the hotels create a feeling of landing at an amusement park. A close second is Tokyo, but not the airport itself. As we set course over the water, we can see literally hundreds of fishing boats with lights blazing to lure the fish to the surface. Flying over Holland provides a view of hundreds of greenhouses, and the border shared by Pakistan and India is completely illuminated at night — an 1800-mile (2897 km) array of lights. There is no limit to the fascinating sights you can see from the air: dimly lit Arctic settlements during the long Arctic night, moonlight dancing across the snow cover or over lakes in the summer, fireworks, oil rigs burning off gas, and the rising moon.

A very welcoming nighttime sight is the array of landing lights at the destination airport. The runway's brilliant approach lights are impressive, their different colors communicating information from the ground to the pilots. Years ago, as a flight instructor teaching landings at night, I asked the control tower to turn up the lights to full brightness — strength five. It proved almost blinding from above; the controller joked about the electricity bill going up.

Colored lights are important markers on the body of an airplane as well. The international standard is a green light on the right-wing tip, red on the left, and white on the tail. Airliners also have white strobe lights to make them easier to spot. However, when we're in a cloud layer, the strobe lights may look like lightning and be a tad annoying, so if the cloud layer persists, we might turn off the strobes until we're out of the cloud.

Another advantage to night flight is the smoothness of air. Much of the turbulence encountered at low levels is due to the sun's heat; at night this abates.

I've flown under an amazing umbrella of stars, and I still marvel at what I see. I have a fond memory from years ago, when I was a first officer transiting the Atlantic Ocean during the wee hours, destined for Europe, and a captain pointed out the astronomical bodies above. He commented not only on their present location, but on where they would be as we crossed eastward into the night. Sadly, I've forgotten most of the discussion. Nowadays, many pilots flash up an app on their phone to converse about the heavenly bodies, making me ask why I've never taken an astronomy course. Next time the cabin lights are dimmed, be sure to look outside.

ARE AIRPLANE CABINS PRESSURIZED TO SEA-LEVEL PRESSURE?

Pressure, temperature, and humidity all decrease with height. At cruising altitude, pressure is only about 25 percent compared to surface pressure, and temperature hovers at a frigid –57°C (–71°F), with relative humidity equivalent to desert air. Most passengers appreciate that an aircraft is pressurized because the density of air and thus oxygen levels decrease with altitude. You know the drill: "In case of a rapid depressurization, oxygen masks will drop in front of you," so start sucking. But many assume the cabin is pressurized to sea level. Not true. In fact, most airliners pressurize to an equivalent pressure height of a mountaintop 8000 feet (2438 m) above sea level. One uniqueness of the B787: the cabin pressure is reduced to the equivalent of 6000 feet (1829 m), lessening the effects of pressurization on long-haul flights. Many falsely believe the cargo holds down below are not conditioned with air, and thus assume it is unpressurized. Not so. That same air that pressurizes the cabin is also piped to the cargo holds, so your pets are safe and the toothpaste tubes in your luggage shouldn't explode; however, the temperatures below on some aircraft can be rather cool. Because of it, some airlines have a blackout period during the year where pets aren't allowed to travel in the belly.

✈ At high cruising levels, the Boeing 787 Dreamliner (and now the new kid on the block, the wide-body Airbus 350) pressurizes the cabin to a denser 6000-foot cabin altitude with the help of fabrication techniques and material (composite versus aluminum), and it even adds humidity so your experience will be more pleasurable during and after a long-haul flight. As well, the advanced B787 does not use air from the engines to pressurize the cabin like other airliners do; instead, it incorporates four electric air compressors. You may notice cabin airflow decreases during engine start as pressurized air is used to start the engines. Yet another uniqueness of the Dreamliner is that the engines are started electrically, but electrical loads are shed (reduced) so the cabin fans still go quiet.

Thin-air facts:

- Altitude affects smell and taste buds, so food preparation and selection are considerations when new food options are introduced.
- Because of the lower pressure at altitude, water does not boil at 100°C (212°F), so your coffee may only be 94°C (201°F) instead.
- Because cabin air is dry, try to stay hydrated with water and juices and if possible consume less diuretics like alcohol and coffee. I know, easier said than done.
- The pressure difference between the outside and inside is about 8 to 9.5 PSI (Pounds per Square Inch).
- Lavatories and galley sinks require a vacuum blower to create a pressure difference while on the ground. If for some reason

the blower is unserviceable, everything works above 16,000 feet (4877 m), where the pressure difference is sufficient.

- Commercial aircraft must maintain a cabin altitude of 8000 feet (2438 m) or less, which means there is about 25 percent less oxygen compared to sea level. This means alcohol consumed at altitude affects you more strongly, so be prudent in your consumption.

✈ I can write numerous accounts of passengers who thought they could handle their booze or decided to indulge with their own stash. Some were met by law enforcement officers or were given stiff warnings. When I was working for a previous airline, one reluctant passenger ripped off the armrest as two burly Boston enforcement officers pulled him out of his seat. It's not worth it, people: when a flight attendant says that's enough, concede.

The pressurization system is designed to raise and lower pressure values within a range for passenger comfort. As soon as the aircraft is rolling down the runway, the system is at work pressurizing. Likewise, when an airplane descends, the computerized system automatically depressurizes the cabin. Gone are the days when an airplane pilot has to adjust the rates. An airliner may descend at a rate of 1500 to 2500 feet (457 to 762 m) per minute, but there is no way a human would find this comfortable, so computers lessen it to about 200 to 300 feet (61 to 91 m) per minute. At touchdown, the cabin pressure is the same as the outside. This change in pressure is harder to handle for people with head colds and for small children. That's why you will hear the odd wailing child letting you know that the aircraft is descending and they are not happy. During ascents and descents, air must be replenished through the eustachian tube in the middle ear cavity to equalize pressure with the cabin. When air is trapped in the middle ear, it can be painful. During ascent, air escapes through the tube easily. Unfortunately, during descents, when pressure in the middle ear must be increased, the eustachian tube does not open so readily. The situation is aggravated with a cold, allergies, or a sore throat. Flight crew take this seriously and will not fly when they have severe head colds. Something to think about if you have one.

To minimize discomfort, you should make a conscious effort to swallow on descent, which will help equalize the pressure in your middle ear. Sucking on a lozenge can help, as can offering little ones something to suck on, like a bottle. Remember it is very dry in the cabin, about 5 to 10 percent humidity for most airliners, which is equivalent to desert air. Dehydration will enter the equation, so stay hydrated. It's why flight attendants pass through the cabin several times on long flights, offering a water service. One final word of advice: make sure that the tops to your toothpaste tubes and shampoo bottles are securely tightened — pressurization has been known to cause the odd lid or top to pop open. This happens even to seasoned aircrew.

HOW IS THE CABIN AIR KEPT FRESH? ARE FILTERS USED?

Air for the cabin is continually bled from the engines. A HEPA (High Efficiency Particulate Air) system filters the air much like filters in hospital surgical rooms. Compared to buildings, however, airliners have even better filtration, a higher air-change rate, and a higher proportion of outside air. Cabin air is generally exchanged every two and a half to three minutes — i.e., flushed 20 to 24 times every hour. Something to think about when you perform a number two in the lavatory.

FLYING FIT.

During flight, and on the ground, many airliners are analyzed by diagnostic software. One such system is AHM (Airplane Health Management), which evaluates thousands of parameters by collecting real-time data on Boeing aircraft and sharing it with maintenance and engineering. The information is so in-depth that a maintenance engineer armed only with a laptop knows how many times the toilets flushed on board. AHM can diagnose an individual aircraft or summarize the health of an entire Boeing fleet. Recently, while taxiing for departure in London Heathrow, we had a maintenance issue that needed addressing. With a quick call via satellite phone, our Canadian-based maintenance went to work diagnosing the airplane. My B787 quickly got a clean bill of health. AHM knows about everything from flight control movements to tire pressures.

AHM reduces delays by sending real-time data from the air to the ground so repair teams can begin to work on a solution before the airplane lands.

WHO IS RESPONSIBLE FOR ROUTE PLANNING?

Numerous desks work 24/7 at flight dispatch centers. Sophisticated software juggles a multitude of factors to come up with the most feasible route. Most routes change on a daily or even hourly basis. Upper winds are near the top of the list of parameters that affect route planning. Among other factors that go into the number crunching are distance, airspace restrictions, inclement weather, altitudes, and aircraft weight. Occasionally, pilots will query the routing and it can be modified if need be.

THE FLIGHT PLAN: A REQUIRED PILOT DOCUMENT DEPICTING ROUTING, ADVISORIES, AND WEATHER.

Over 700 flight plans, consisting of three sections, are generated daily for my airline. (An equal number of flight plans is created for the connectors at a different dispatch office.) The first section depicts the most feasible route, with winds and weather being major players. The second section, NOTAMs (Notices to Airmen), advises the pilot of possible taxiway or runway closures and navigation and approach aid outages, to name a couple of items. Weather is the last section, depicting en route weather, upper winds, areas of turbulence, jet stream locations, airport forecasts, and present weather. A domestic flight plan is about 25 pages long, whereas a 14-hour long-haul flight may require 50 to 60 pages. That's a lot of reading!

Planning facts:

- Even though direct contact does not exist between the pilot and dispatcher, we can communicate by phone, datalink, VHF radio, and sometimes satellite phone to discuss the flight plan.
- More and more aircraft can upload the flight plan directly into the flight computers via datalink.
- One dispatcher may generate 40 flight plans per shift. A flight plan is generated about two to three hours before departure.
- Next to the pilot group, flight dispatchers are the second-highest paid, and there is a present demand. Something to think about if you love aviation but prefer to sit on terra firma.

THE SCOOP ON GREAT CIRCLE ROUTES. HOW DOES FLYING IN WIDE ARCS, INSTEAD OF "AS THE CROW FLIES," REDUCE TRAVEL TIME?

We indeed try to set course "as the crow flies." This is the shortest distance on a sphere like Earth, but its depiction on a flat map is often distorted. One would think a Chicago to Hong Kong flight would entail flying over California and the Pacific Ocean, but the shortest distance by far is over the North Pole. These curved or hyperbolic arcs, called "great circles," are truly the shortest distance between two airports over planet Earth. Even before Christopher Columbus sailed the ocean blue in 1492, mathematicians knew about great circles and navigation.

But because of airspace restrictions, airspace closure, military active areas, no-fly zones due to wars, and political restrictions, we don't always fly a great circle route.

Here's an example: the great circle distance from Toronto to Hong Kong is 7810 miles (12,570 km). If I calculated that route from Toronto to Hong Kong via San Francisco, it would be 9186 miles (14,780 km). On the return flight, we tend to fly a more southerly route to capitalize on tailwinds.

The sliced-orange analogy is overused, but it's a good one. The following explanation is borrowed from the internet. Take a pen, mark two dots on an orange (not on the "equator" or straight up and down along a "meridian" — these are already "great circles"), and then slice the orange between the two dots, with the knife angled toward the very center of the orange. If you remove the peel in one piece and flatten it, what appeared to be a straight line on the orange becomes a curved line on the peel. So the closest distance between two spots on a sphere is plotted as a long curve when using a flat representation. I tried this with a tangerine — I had no nearby oranges. I couldn't convince myself, so I ate the experiment instead.

> ✈ On a recent flight from New Delhi back to Toronto,
> we had to deviate around Pakistan's closed airspace as
> major friction existed between the two countries. This
> "pissing contest" meant our flight required another 900
> miles (1450 km) of navigation, and of course extra fuel.
> Thus we had to land in Denmark and lay over, leaving
> passengers wondering why the deviation. Good ole
> politics can cost airlines millions. As I write this, Hong
> Kong and mainland China are playing silly bugger,
> canceling hundreds of flights. The CEO of Cathay
> Pacific airlines and other executives have resigned.

GETTING AROUND: JUST HOW DO PILOTS GET FROM AIRPORT A TO AIRPORT B?

Back in the day, pilots flew referencing ground-based navigation devices called beacons and VORs (VHF [Very High Frequency] Omnidirectional Range). These devices still exist but are giving way to GPS (Global Positioning System). Now we fly mostly from dot to dot — specific points in space called waypoints — allowing more efficient and more abundant routes. These waypoints can be close together, within a few miles near airports, or spread hundreds of miles apart over the oceans or in sparsely settled areas. Gone are the days when a dedicated navigator tuned and identified grounded navigation sites. The names of waypoints may have some local importance, but most are spewed from a computer with a combination of five letters. These uppercase five-letter combos usually have a vowel, but not always.

As I write this, I am sitting in the hotel lobby in Santiago, Chile. Another crew is flying a shuttle to Buenos Aires and back. I will continue the flight to Toronto during the wee hours of the morning, touching down as the sun comes up. Their flight will follow the dotted airway of ASADA, IRASU, BUTMO, GPI (General Pico), LOLAS, ANKON, and MUPAV to an arrival into SCEL (Santiago), a.k.a. Arturo Merino Benítez International Airport, named after an aviator with commodore status.

My favorite waypoint — and to be honest, I am surprised it still exists, considering political and gender correctness — is CZI (Crazy Woman) in Wyoming. I still chuckle when cleared to that waypoint. I imagine the controllers working that sector receive a lot of snide remarks. You can find obvious humor in some running combinations. ITAWT, ITAWA, PUDYE, TTATT, IDEED is the approach into a smaller airport in Portsmouth, New Hampshire. The translation requires having watched the Bugs Bunny cartoons, when Tweety bird sees Sylvester the cat and says, "I tawt I taw a puddy tat, I did!" It may take a bit, but you'll get it.

The DYAMD (Diamond) approach into San Francisco, which I have flown over 120 times, has this sequence of waypoints: DYAMD, LAANE, ALWYS, FLOWZ, FRELY — i.e., the diamond lane (on a highway) always flows freely. Who knew?

When flying into Toronto Pearson's Runway 23, the name of the approach (one of 18) is the IMEBA. We fly to IMEBA, then LEPUX, DUGDA, CALVY, VEPVU, OMTOK to the runway. These generic computer-generated names supposedly have no language preference, so a Russian Aeroflot pilot should be able to pronounce them as easily as a United Airlines pilot; however, I find myself stuttering on a few. Can you imagine challenging a commercial pilot to a game of Scrabble using waypoint names? We still use latitude and longitude coordinates, particularly over sparsely settled areas and especially over the oceans. While flying across the Atlantic (colloquially known as the Pond), we may fly to 50N 50W, which translates to 50° north and 50° west. Waypoints like these must be confirmed in our navigation computers, followed by a heading and distance check to eliminate

GNE (Gross Navigational Errors). That is the last thing I want to hear from air traffic control: "Confirm your heading?" It's one of those queries an airline pilot has nightmares about. But lately, our flight plan directly uploads into our flight management computers. Some of our long-haul flight plans can be five to seven pages of waypoints, and up to about a few years ago, would require painstakingly inputting the points and airways. Not anymore . . . unless the datalink is not working. Think of datalink as text messaging via radios and now satellites.

DOES THE EARTH'S ROTATION SHORTEN YOUR TRAVEL TIME?

Our routing is affected by the Coriolis effect — but not because the Earth turns underneath us. The Coriolis phenomenon deflects everything to the right in the northern hemisphere (left in the southern), but our sophisticated navigation systems factor it in. What really affects our time en route are high-altitude winds, which can blow as strong as 230 knots (265 mph or 426 km/h) and tend to prevail from the west.

ARE THERE TWO NORTH POLES?

Yes, there is a magnetic north and a true north. The true North Pole is where our Earth rotates about its axis and where Santa Claus is rumored to reside. Latitude and longitude, lines of navigation, reference true north. But headings in navigation reference magnetic north using a compass, with its location deviating from true north. A compass heading of north points to magnetic north, presently about 200 miles (320 km) from true north. Because aviation uses both North Poles, variations are factored in. However, when we fly "over-the-top" polar routes, we only fly by true north.

OVER-THE-TOP POLAR FLIGHTS.

Santa Claus is not the only one flying in North Pole's airspace. Many airlines have been flying "over the top" since the early 2000s. But sometimes because of solar activity and extreme cold temperatures, polar routes are not optimized.

Polar facts:

- Popularity has increased the number of designated polar routes from four to ten.
- The closest distance a polar route comes to the North Pole is 60 miles (97 km).
- It takes six hours to fly to the North Pole from Chicago, New York or Toronto.
- Each polar flight is equipped with two polar kits containing winter clothing in case of a flight diversion. These kits include: a yellow parka, boots, mittens, knit face mask, winter hats, and gloves.
- No flight can travel through temperatures below –65°C (–85°F) for 90 minutes or more. Think possible fuel congealing.
- Solar flares and cosmic radiation are closely monitored, and if guidelines are exceeded, no polar routes are flown.
- Polar flights tend to be smoother, as the jet streams corkscrew around the globe much farther south.
- True north is referenced instead of magnetic north because the proximity of the magnetic pole causes erroneous readings. Navigational displays must be switched from magnetic reference to true.
- Magnetic north, located in northern Canada, is moving northwest at about 40 miles (64 km) a year and will soon

enter Russia. But lately it is speeding up while heading to Russia. Runways named after their magnetic heading will change over time. I soloed on runway 24 (238°) in Halifax 42 years ago, now it's called runway 23 (233°). (You know you are getting old when runway nomenclature changes.)

- The weatherman's winds are given in true north headings, but when a pilot lands or takes off, magnetic wind direction is used.
- Advantages of polar flights: reduced flight time, no en route fuel stops, new routes allowed, absence of turbulence, and shortened aircrew duty time.

SANTA THE AVIATOR (CHRISTMAS 2004).

According to the onboard navigation computers, the aircraft was precisely over the North Pole, signifying Santa Claus and his workshop had to be directly below! Unlimited visibility prevailed in winter's starlit darkness on a recent "over-the-top" non-stop flight from Toronto to Hong Kong, offering a rare glimpse of the North Pole as I scanned the frozen terrain for hints to Santa's whereabouts. Though my last pilot medical still deemed my eyesight faultless, it was uncertain whether I spotted the glowing lights to Santa's busy toy factory with elves working frantically inside. On returning home, my six-year-old son at the time was ecstatic to hear of my bird's-eye view. During my search, I pondered the similarities and differences between an airliner and Santa the aviator. It's a known fact Santa Claus prefers a snow-covered landing pad, but not so for airplanes. Pilots prefer runways to be bare, and if dubious conditions exist, ground equipment will leap into action incorporating snowplows, sweepers, and vehicles to spread non-corrosive chemicals to melt snow and ice. We also consult charts to determine if the reported braking action and crosswinds are within limits; if not, it's off to another runway or airport for landing. It's well known that Santa lands on rooftops with very steep pitches.

Our takeoff and landing must be within 2° of slope. Very sophisticated landing instruments both on board and at the airport are a must for us to locate the runway in snowy conditions. I'm certain Santa doesn't navigate using GPS or laser-mounted gyros to detect momentum shifts . . . or does he navigate by "dead reckoning" from a magnetic compass, labelled ineffective in extreme northern latitudes?

When we traverse the North Pole, our airspeed is about 86 percent the speed of sound. One website armed with statistics on Santa Claus postulated that his required speed must be 3000 times the speed of sound to allow him to reach everyone on Christmas Eve. He must also have special permission to bust the mandatory airspeed restriction of 250 knots (288 mph or 463 km/h) below 10,000 feet (3048 m) above sea level. Rules regarding appropriate navigation lights must also be twisted, although Rudolph's red nose could improvise as a red anti-collision light. There's also deicing systems required as he enters cloud, and the gamut of instruments necessary to keep the sled upright in disorienting cloud. (A non-trained person flying in cloud is statistically proven to last under a minute before plunging into a spiral dive.) The FAA or Transport Canada could easily ground Santa's sled on hundreds of violations, but in the name of "Christmas spirit" Santa has been given a special flight permit.

As you know, to stay current, airline pilots must undergo rigorous testing every six to eight months in a flight simulator. For the pilot, one of the toughest hurdles to nail down during a "check ride" (flight test) is the loss of an engine on takeoff. This reminds me of a story which circulates in the aviation world about Santa's "check ride." Apparently, even Santa Claus could not escape the required flight test with a flight inspector. In preparation, he had the elves wash the sled and bathe all the reindeer. Santa got his logbook out and made sure all his paperwork was in order. He knew the inspector would examine all his equipment and truly put his flying skills to the test. The examiner walked slowly around the sled. He checked the reindeer harnesses, the landing gear, and even Rudolph's red nose. He painstakingly reviewed Santa's weight and balance calculations for the sled's enormous payload. Finally, they were ready for the check ride. Santa got in and fastened his seat belt and shoulder harness and checked the compass. Then the examiner hopped in carrying, to Santa's surprise, a shotgun. "What's that for?!?" asked Santa in disbelief. The examiner winked and said, "I'm not supposed to tell you this ahead of time." Then he leaned over to whisper in Santa's ear, "But you're gonna lose an engine on takeoff."

All airliners around the world are equipped with devices to interrogate other aircraft in close proximity, determining direction and altitude, and give a resolution advisory if need be. It's not known whether Santa's sled has been updated with this new technology, so it will be hard to see him coming. However, having flown many Christmas Eves, we generally make a passenger announcement advising passengers that air traffic control has detected an unknown blip on their radar originating from the North Pole.

One Christmas Eve, I flew from Calgary to Frankfurt (junior crews both in the flight deck and cabin will have the distinction of working through Christmas). Our flight plan took us over Baffin Island and Greenland . . . two great vantage points to spot Santa and his hard-working reindeer. If you're on my flight, expect a briefing on Santa's whereabouts and a "season's greeting" from the flight deck.

✈ Christmas of 2019, I had the honor of flying a red-eye from Vancouver to Toronto while others slept through the night awaiting Santa. I mentioned in my welcome-aboard PA that air traffic control had spotted Santa on the radar and we would be racing him to Toronto. Got a few chuckles. My son, who just turned 22, followed me on this pairing as a passenger. Santa can scoot . . . as well as time.

AVIATE, NAVIGATE, COMMUNICATE — BUT RADIATE?

The cardinal rule for any pilot is to aviate, navigate, communicate . . . but why a reference to radiation? Cosmic radiation is predicated on four factors, and it's part of doing business. But before you think you will be glowing in the dark, it turns out we are at less risk than that! The four factors of solar radiation are: latitude, altitude, flight duration, and solar activity. I am a meteorologist, but this topic is out of my field of expertise and much of it is derived from a recent pilot union survey. The term "cosmic rays" refers to elementary particles (such as electrons and protons), nuclei (such as helium nuclei, or alpha particles), and electromagnetic radiation (x-rays and gamma rays) of extra-terrestrial origin. These cosmic ray particles are produced by high-energy events very distant from the solar system, such as supernova explosions. They come from all directions in space.

Aircraft at higher altitudes receive greater amounts of cosmic rays than at lower altitudes. But jet engines like higher altitudes — in fact, the rule is generally the higher the better. When I'm at flight level 350 (35,000 feet, or 10,668 m, above sea level) and I look up and see a business jet at 47,000 feet (14,326 m) above sea level, I think of their possible dosages.

The north and south polar regions are less shielded by the Earth's magnetic field than regions near the equator. The Earth's magnetic field shields us from cosmic radiation. Remember, we are talking magnetic North Pole and not the "true" North Pole. Presently, we in Canada must contend with the disparity of protection, with ours being more than Ivan in Siberia — for now.

The sun's activity varies over an 11-year cycle. When the sun is active, it shields the inner solar system from cosmic radiation. When the sun is inactive, Earth receives more cosmic radiation! Like most people, I would have thought the entire opposite. When sunspots explode, they hurl massive clouds of hot gas away from the sun. These clouds contain not only gas but also magnetic force fields. Magnetic fields deflect charged particles, so when these clouds sweep past Earth, they also sweep away many of the electrically charged cosmic rays that would otherwise strike our planet.

The dose from radiation that a person receives is measured in units of sieverts (Sv). The average person receives about 2.6 millisieverts (or 0.0026 Sv) per year from normal natural, industrial, and medical sources. The normal range is from one to four millisieverts. A typical medical x-ray, for example, generates a dose of about 0.01 to 0.1 millisieverts. A typical transcontinental flight will generate a dose of about 35 microsieverts (.035 millisieverts or 0.000035 sieverts). A microsievert (μSv) is 1000 times smaller than a millisievert (mSv). Thus it would take over 75 such flights per year to generate a received dose that is roughly equivalent to the normal annual background dose.

Recently, the company PCAire (Predictive Code for Aircrew Radiation Exposure) conducted a radiation exposure analysis on myself and my colleagues. After all the facts were compiled, we remained below the recommended intervention level of 6 mSv, which is one-third the limit of 20 mSv used for occupational workers. For pregnant aircrew, the dosage should be no more than one mSv annually. Many would assume the B777s flying transpolar flights would see the higher doses. Not so, because they usually fly at lower altitudes, where the exposure is less. When they become light enough to fly higher, they are heading south over Siberia, so the exposure gets less and less.

✈ Even though the "A" in "PCAire" stands for "aircrew," a frequent flyer can also log into the website and get dosages for each flight. My 2019 dosage was 4.8 mSv.

Doug building flight time in his early years.

Doug as a new pilot at Air Canada.

TOP: *The flight deck of the B787.*

LEFT: *B787 simulator at a training center in Toronto.*

BELOW: *Flight dispatch in Toronto.*

B787's left engine viewed from the washroom window.

Getting deiced in Montreal.

PHOTO BY BRIAN LOSITO

DIAGRAM BY CAERINA ABRENICA

The colloquial four phases of a jet engine: suck, squeeze, bang, and blow.

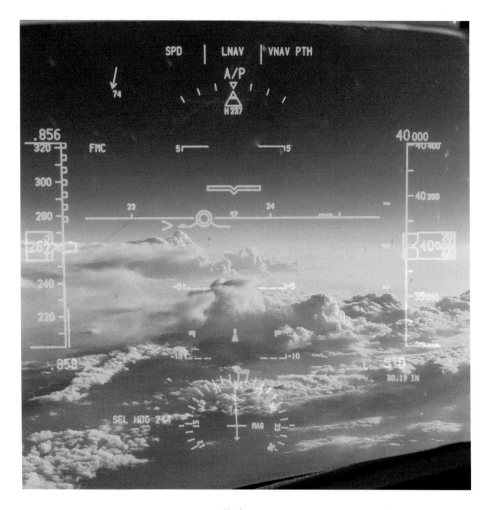

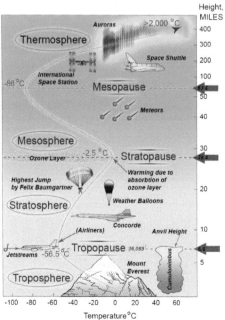

ABOVE: *Weather up ahead as seen through the HUD (Head-Up Display).*

LEFT: *Depiction of the standard atmosphere.*

NORTH ATLANTIC TRACKS — TRANSATLANTIC FLIGHT ROUTES.

Some 2500 flights cross the North Atlantic Ocean daily, a mind-boggling rise in transatlantic air traffic since the first solo flight by Charles Lindbergh, in the *Spirit of St. Louis* in 1927 from Long Island, New York, to Le Bourget, Paris. To ensure safe passage between northeast North America and Western Europe, routing instructions called North Atlantic Tracks (NATs) are issued twice a day by area control centers in Gander, Newfoundland, and Northern Ireland. Like transatlantic superhighways, NATs keep airliners evenly spaced at 10 minutes, with each track 60 nautical miles (one degree) apart, and steer them to take advantage of fuel-saving tailwinds or clear of delay-causing headwinds. Recently many routes have shrunk to 25-nautical-mile separation to accommodate aviation's ever-increasing traffic and because of airliners' proven accuracy. Realistically the separation has dwindled to half a degree (30 nautical miles).

Interesting enough, the SST Concorde got above all the tracks' highest altitudes and flew its own route, adding to its uniqueness.

✈ It's January 1, 2020 (happy new year!), as I write this, and a new air traffic control system (ADS-B) became mandatory today in the United States. This system has further reduced separation over the Atlantic to a mere 15 miles (24 km).

Shanwick is the air traffic control name given to the area of international airspace above the northeast part of the North Atlantic. The name is a portmanteau of two air traffic facilities concerned with flights operating within this airspace, these being the Ballygirreen Radio

Station, six miles (10 km) North of SHANnon Airport in County Clare, Ireland, and the PrestWICK Centre in Ayrshire, Scotland.

There are generally five to eight NATs (sometimes more), each designated by a letter. Eastbound nighttime NATs start at Z (southernmost route), followed by Y, X, W, to as far as N (if need be) but omit O. Westbound daytime NATs are A (northernmost track), B, C, D, as far as M, omitting L.

EYES IN THE SKY — NEW SATELLITES GIVE GLOBAL AIRCRAFT SURVEILLANCE 20/20 VISION IN 2020.

Global air traffic will soon be watched from above, under a constellation of 66 LEO (Low Earth Orbit) satellites. Ground-based surveillance radar is limited to line of sight; it is a challenge in mountainous regions and nonexistent over oceans. The global umbrella network of ADS–B (Automatic Dependent Surveillance–Broadcast) satellites will safely allow aircraft to fly closer to one another (15 miles over the ocean, down from 30 to 60) and permit shorter routes and optimal altitudes, thus saving fuel and time and lowering carbon emissions. Many countries are already using the system. Aircraft flying in U.S. airspace are required to be ADS–B compliant on New Year's Day 2020, with Canada following in 2021 and 2022.

Before this innovative system, 70 percent of the globe, including polar regions, oceans, mountainous regions, and even deserts and jungles, had no access to surveillance. Many are surprised by the lack of radar coverage in sparsely settled areas. One flight that brought to the forefront the gaps in radar coverage is Malaysia Airlines Flight 370, the disappearance of which is now the biggest mystery in aviation.

WHAT IS THAT RUMBLING HEARD AT CRUISE PHASE?

Generally, the higher the altitude the more efficiently the engines burn fuel. If the aircraft is heavy, the pilots must wait until enough fuel is burned to lighten the aircraft, enabling the aircraft to climb higher. This is called a step climb. For a large airliner flying a long-haul flight, this step procedure can take hours before the pilot can reach maximum altitude. I've flown flights where it took 10 hours to get to our final cruising altitude via several step climbs. But pilots may also climb to find smooth air. We don't always stay at the same altitude, so a slight rumbling may be heard. This rumbling when the engines spool up (accelerate to the proper RPM) tends to be more noticeable abeam (off to the side) and aft of the engines.

HODGEPODGE OF AVIATION UNITS.

Whether a pilot is flying to Milan, Brazil, or Hong Kong, English is the international language of aviation and the predominant language used. Don't expect a controller in Bucharest to know German for a Lufthansa pilot, but they must know a minimum of air traffic control English. But you'll hear Lufthansa pilots speaking both German and English in Germany. The commonality is English, so why is there not a similar consistency for units of measure? Probably for the same reason that we drive on the right side of the road in North America but on the left in England, or that Celsius rules the planet but Fahrenheit presides in America.

One would think units would be the same around the world, especially in light of safety. But they are knot — oops, I meant *not.* In a car we measure speed in either miles or kilometers per hour, but not in an airplane. The knot stems from marine use (charting and navigating). Driving down the highway at 60 miles per hour means you are covering 60 statute miles (5280 feet each) every hour. But 60 knots means 60 nautical miles (6076 feet each) per hour. International pilots are constantly shifting from one system to another using a mix of metric-derived SI (International System of Units from the French *Système international*), imperial measurements, and naval standards just to make things interesting. Fortunately, the push of a button in modern flight decks converts many units such as pressure, altitude, and fuel.

Here is the hodgepodge list:

- Altitude is in feet except over China, Mongolia, and North Korea, where it's meters. Russia uses feet at higher altitudes but meters at lower heights. Cloud heights are all in feet.
- Celsius for temperature, but the American public converses in Fahrenheit.
- For pressure, inches of mercury are used for a pilot's altimeter to fine-tune height above sea level in North America, but hectopascals are used elsewhere. PSI (Pounds per Square Inch) is used for aircraft tire pressure and for aircraft systems such as pressurization and hydraulics.
- Weather charts use millibars in the United States but hectopascals worldwide.
- Visibility is measured in statute miles and feet in North America, but kilometers and meters elsewhere. This is a bit challenging when visibilities become low.
- Winds are measured in knots, but not in China, Russia, and Mongolia, where they are meters per second. Rule of thumb: double the meters per second to get knots — e.g., 10 meters per second is 20 knots.
- Fuel is measured in liters, imperial gallons, and U.S. gallons.
- Weights are in either pounds, tons, kilograms, or tonnes (metric tons — 1000 kilograms).
- Speed is knots, but at higher altitudes Mach is used. A Mach, given as a decimal fraction, is the aircraft's true airspeed divided by the speed of sound. More about this unique unit later.

HOW CLOSE DO OTHER AIRPLANES GET?
TCAS (TRAFFIC ALERT AND COLLISION AVOIDANCE SYSTEM).

Airplanes crisscross one another's paths more frequently than you think. Separation must be 1000 feet (305 m) vertically and about three to five miles laterally. During the late '80s, a system was built so each aircraft could electronically interrogate other aircraft. If a conflict arises, the TCAS will command one aircraft to maneuver one way and the opposing aircraft to fly another route. There are two levels: a TA (Traffic Advisory), meaning an airplane is getting close but poses no immediate threat, and RA (Resolution Advisory), when pilots must take evasive action. This procedure is one of the many things practiced in the simulator.

WHEN YOU MAKE A PA (PUBLIC ANNOUNCEMENT),
DO YOU READ FROM A SCRIPT?

Most pilots ad-lib, and we all have our unique style. Some pilots who learned another language may take out their "cheat sheets." I remember one pilot impressed a full load of passengers bound for Tokyo with his Japanese, learned at night school. Thirty years ago, I took a six-week French immersion course, so I too read from a script, but now I let the flight attendants translate because my "pilot French" is now passé. Flight attendants have a small booklet they can reference but in-charges have iPads for PA references. Numerous studies reveal passengers appreciate the "welcome aboard" from the flight deck. But sometimes the work to get us out on time pre-empts it. Surveys confirm that passengers want the facts, appreciate updates, and like them personalized, so I include my name and those of the pilot(s) I am flying with. On smaller aircrafts, some pilots include the flight attendants' names. Pilots may state, "In the cabin, you have Michael, Nancy, and Mario." I make my PA as if I am addressing the passenger seated in 22B who possibly flies once a year: I tone things down to alleviate concern and ensure understanding. Most won't understand "it is warm in the cabin because the APU is U/S" but will understand "the air conditioning unit is not working today."

Oration facts:

- Terms some pilots assume passengers know: APU, ATC, hold, light chop, wheels up time. Translation: APU — Auxiliary Power Unit, literally a small jet engine in the tail that conditions air or supplies electricity; ATC — air traffic control; hold — holding pattern; light chop — turbulence; wheels up time — when we must be airborne due to air traffic constraints.

- When teaching weather to the new-hire pilot class, I tell them they should try not to include certain words. For example: "Ladies and gentlemen, we will be 15 minutes late arriving into Los Angeles because we are 'bucking' headwinds.'" The word "bucking" may sound like something else over the PA.

- Recently, my first officer gave our arrival time and weather, stating the clouds were broken. "Broken" is pilot talk denoting cloud covering most of the sky. A deplaning passenger fretted because we said something was "broken."

- Most Canadian pilots are not bilingual, so we rely on the flight attendants to translate. I am amazed how gifted they are at translating, especially with sometimes complicated announcements. Having said that, I am told some flight attendants do not like translating when the pilot's announcement is longwinded.

- When I taught new hires the ropes, I could see many were apprehensive about making PAs. To alleviate things, I told them to write things down prior to their announcement.

It was advice I received years ago when I was in their shoes.

- It takes finesse and practice to make announcements to a full load of passengers. Most pilots are great at it, but others need some honing or they don't do it at all.
- Speak clearly, slowly, confidently, a tad louder than normal, and avoid complicated jargon.
- It is imperative a PA be made if we must deice the aircraft. It's the first thing on the deice checklist.
- Thunderstorms may be deemed "heavy showers" or "inclement weather" to alleviate concern. Mentioning pending turbulence is a judgment call; some truly appreciate the heads-up but others prefer not to know.

PA AND RADIO GAFFES.

Sometimes, and this never happened to me, pilots push the wrong button and make the "PA of all PAs." Instead of serenading the passengers in the back, their PA is transmitted over the air traffic control frequency. Over the years, some long-winded ones have been misdirected over the airwaves. Usually, a brief silence ensues, and then numerous pilots chime in: "Nice PA, when are we landing?" "Wrong frequency," etc., causing the pilot to sink further into their seat after realizing their gaffe. Some pilots really caused grief by complaining about air traffic control to their passengers when in fact their plaints boomed over the airwaves.

And then there is the "hot mike," where the microphone-transmit button sticks open and transmits everything. Immediately, you will hear pilots saying "check your mike" on another audible frequency. This scenario can jam a frequency, so pilots are told to go to another frequency. It also happens in air traffic control, and I've heard some interesting comments over the years from controllers not knowing their conversation was live.

Then, of course, there's transmitting on the wrong frequency. Instead of communicating with company (we have one radio for air traffic control, and the other is for company or monitoring an emergency frequency), the transmissions boom out over the emergency frequency. Immediately you will get the "guard police" (other pilots taking it upon themselves to correct things) telling the gaffing pilot to "check your freq" or "you're on guuaarrd!!!" Guard is the emergency frequency.

✈ "Peeing the mike." As I write this, I am deadheading — yet again. The captain made two great announcements; however, he peed the mike. Saying the letter "P" involves the puffing of air. Because of this, many words are distorted. It irritates me when I hear pilots peeing their transmissions. But how do I tell that to a seasoned B777 captain? "Great announcements, but you peed the mike." All it takes is to move the microphone away from their mouth a little. You'll find many pilots, including me, moving the boom mike accordingly to avoid this annoyance. Similarly, I find it annoying when pilots speak into the mike when they're near their flight deck air gaspers, so air is blowing across their mike. Or have their mike too far away from their mouth and then wonder why ATC can't make them out. Air traffic control, after hearing enough (or not hearing), will remind pilots of this. On a very recent flight, Cleveland center mentioned that it sounded like the pilot's mike was in first class and to move it a little closer.

And speaking of clearer transmissions, female voices transmit better than males' over the radios. A male's deeper, lower voice can sometimes sound muffled. I notice more and more female voices heard from air traffic control centers. It's been found, and I notice it too, a female voice sounds clearer over the radios. Maybe it's why a female voice is used for F-18 Hornet fighter jets. It's the voice of a real person recently retired from Boeing. The voice is known as "Bitchin' Betty," and a pilot is more likely to listen. But most airliners utilize a male voice — some call him "Barking Bob."

ON THE JOB: THE UNSUNG HEROES (FLIGHT ATTENDANTS).

An hour or so before takeoff, the flight attendants are already busy preparing for the flight. They review safety and emergency procedures, check all the emergency equipment, and verify that the necessary supplies and commissary have been boarded. A list of the equipment checked is taken to the flight deck, with every name working the flight included. Each of the attendants is assigned a specific location. The B787 requires up to eight flight attendants. The B777 needs 12, whereas a small Airbus requires three. Flight attendants are qualified to work on the entire fleet, whereas pilots fly only one type of aircraft. The number of flight attendants per aircraft type varies according to the company, but there is a minimum each aircraft must have for safety reasons.

Then they go into the "greeting mode" as passengers board, and ensure baggage is stored. Many passengers assume flight attendants are there to heave their heavy luggage into the bin. Sometimes that can make for interesting conversations. They are also tasked to supervise UMs (Unaccompanied Minors), the physically and mentally challenged, the elderly, and any others who require extra help. What a task! The service director — a.k.a. in-charge, "Queen Bee," or grand pooh-bah — will deliver the final passenger count to the flight deck and let the pilots know the cabin is secure: passengers are seated, luggage is stored, and egos are catered to.

During pushback from the gate and taxi to the runway, flight attendants used to stand in the aisles and demonstrate safety procedures. Canned safety videos spare them that now, but they still must stand there in case someone has a question. When I fly as a passenger, I have yet to see someone put their hand up and ask. However, you will still see a live demonstration if the video system goes kaput.

One minute before takeoff, an announcement is given from the flight deck: "Flight attendants, take position for takeoff." Some airlines use a chime system. At this point, flight attendant duties momentarily subside, whereas the pilot's duties escalate in preparation for imminent flight. As the airplane barrels down the runway, most flight attendants are silently reviewing their emergency drills — the "what-ifs."

Definition of a flight attendant: a problem solver, medical attendant, travel guide, translator, mind reader, listener, plumber, cleaner, vomit responder, safety officer, comedian, diplomat, cook, bartender, consoler, teacher, guidance counselor, model, go-between, babysitter, and ego booster/deflator for a pilot's landing. So try not to demean their position by saying "stewardess" or "steward."

> ✈ The titles "stewardess" and "steward" changed to the generic "flight attendant" in the 1970s, but you'll still hear passengers dating themselves.

If you want to know the true character of a famous actor, politician, sports celebrity, etc., then ask a flight attendant. They can tell you about the polite and friendly ones, and about those who have hissy fits, are vulgar, pompous, or just nasty. And the answers may shock you. Some of Hollywood's prim and proper have potty mouths that would embarrass a sailor, and yet some of the rough-and-tough athletes can be the biggest marshmallows on the flight. I know I've looked at many celebrities differently after getting a briefing from the flight attendant.

Without me naming specific flights, flight attendants can tell you which have the friendliest, the most demanding passengers, the most religious, the whiniest, the biggest drinkers, the most laid-back, the politest, the cleanest, the most "special" (think emotional support animals), and which flights require the most wheelchairs. Not sure why there aren't more books written from a flight attendant's perspective. Sometimes I truly wonder how they tolerate and put up with the "against the grain" types and their shenanigans. Put people in an aluminum (composite) tube for hours on end and watch their true colors effervesce.

The most prevalent question flight attendants hear during the flight is "Where are we?" Luckily a moving map display now keeps that question to a minimum. It's funny how passengers assume flight attendants have a built-in GPS. They may have not flown this route in years, plus routes change daily. Sometimes when I am asked where we are, it can be challenging to answer. Sure, I could tell you we just passed a navigation waypoint affixed with a unique name, but it can be a scramble if you want to hear a geographical point. Even the second-most popular question — "How much longer?" — is answered by the moving map display. When our airtime (wheels up to wheels down) is given during our welcome-aboard PAs, some frequent flyers start a timer on their watch or phone when the wheels leave the runway to know precisely when the wheels will touch. Yes, they are watching.

As the flight descends through 10,000 feet (3048 m) — about 10 minutes before landing — the seat belt sign illuminates, signaling it's time to pick up any remaining trash and collect headsets. The use of disposable headsets is also waning, as many passengers bring their own. Visit a gym or observe people walking; nearly 90 percent of people are now plugged in. And people wonder why pedestrian accidents are on the rise?

When the cabin door opens, flight attendants are there, bidding passengers goodbye. If the flight involves a layover, they will be off to a well-deserved rest, and tomorrow these unsung heroes will do it all over again.

In my opinion, one of the toughest jobs out there is flight attendant. You must be courteous, groomed, perky at five a.m., know safety and first aid, handle a business class passenger who won't look at you in the eye, and deal with possibly the most difficult passenger there is — the pilot!

Another prerequisite bestowed upon a flight attendant is bombarding passengers with announcements. I detested this when I commuted. Blah, blah, blah. We must keep certain authorities happy, so it's imperative we tell you how to do up your seat belt, where the exits are, how to work the oxygen masks, and if we are flying over water, how the life jackets work. As a note, search the internet for a picture showing the passengers standing on the wing when Captain Sully ditched into the Hudson River. Virtually no one is wearing that infamous yellow flotation device.

✈ Three sides to a story . . . "There is a passenger in the back of the airplane who just swore at a flight attendant, is angry, and won't listen." I asked the in-charge, "Do you want him removed?" I also asked my first officer to call operations to get some backup. I made it perfectly clear I was behind her decision to deplane the passenger. After all, it would be the four flight attendants who would have to contend with the unruly passenger. It was decided. We would deplane the so-called troublemaker — notice I said "so-called." Luckily, he did not have checked baggage, or it would translate into a 20- to 25-minute delay to find and remove his bags from the belly.

I went back to the front galley to see him off. This fella was about six foot four, and if he and I were in a fight there would be two hits, him hitting me and me hitting the floor. I whispered under my breath to the in-charge as he neared the exit, "This guy is big, I better get my first officer." Then we had a chat in the front of a fully loaded A320. He apologized profusely. He had not been drinking, and the moment he realized this was serious, he turned into a puppy dog. With the concurrence of the in-charge, plus the flight attendant in question and the two dispatched agents, we agreed to give him a second chance. The flight was uneventful.

I talked to the passenger on the bridge in Vegas. He said, "Captain, I wish I could have told my story. It did not happen like that." It turned out his wife was a flight attendant for another airline, and his son was learning to be a pilot at a flight college. I'm wise enough to realize two things from experiences like these. One: the situation could have been handled differently by all parties. Two: there are three sides to a story — your side, my side, and the truth!

→ How to annoy flight attendants: by going to the washroom when they are in the aisle with the food- or drink-laden trolley. It's tough maneuvering these heavy aluminum 200- to 250-pound carts. Sure, you may be greeted by a smile, but you just pissed them off. Another no-no is asking to have your food tray removed immediately after eating. Remember, they are navigating through a jam-packed cabin, so cut them some slack. As I write this, I am on reposition flight from Los Angeles as a passenger. A rotund business class passenger is ripping out deafening snores after being fed and watered. It reminds me of another repositioning flight where I witnessed a sleep-deprived passenger sprinkling water onto the face of a businessman in a desperate attempt to stop his deafening snoring.

✈ During my captain upgrade, I was forewarned about the great and not-so-great in-charge flight attendants. After 15 years as captain, I know there are some who could put out a fire in the back cabin without me knowing it, and then some who will declare a near-emergency because we lack business class drinking glasses.

Also, beware when walking through the cabin after meal service when the cabin is inert with sleeping passengers. The smell from farting passengers unaware of their expulsions may gag you. Flight attendants are briefed on this flight flatulence during training. And we worry about cow farts and burps causing global warming? The global warming academia should take a long-haul flight. Even though we have HEPA (High Efficiency Particulate Air) filters, the cabin air can be foul.

✈ One flight attendant's embarrassing moment: she noticed a business class male passenger making rhythmic up and down movements under his blanket. She couldn't believe such blatant maneuvers in public. After seeing enough, she confronted the man, only to find he was cleaning his glasses.

> ✈ Flight attendant jokes about pilots: What does a pilot use for birth control during a layover? His personality, or his layover clothes. How can you tell the airline pilot at a cocktail party? You don't have to, they'll tell you.

OTHER UNSUNG HEROES: RAMP ATTENDANTS, COLLOQUIALLY KNOWN AS "RAMPIES" OR "RAMP RATS."

These people work in the extreme spectrum of weather. Peering out the flight deck window, I see them bundled in layers to combat the cold or sweltering in the inhospitable heat that permeates the belly of the airplane. My hat goes off to them as well as to the fuelers, commissary, and maintenance — all of who must contend with Mother Nature while I sit cozy in a white short-sleeve shirt.

> ✈ During an après-game debriefing in the hockey locker room (think beer) with mostly "rampies," I was told this amusing anecdote. One flight was transporting several bars of gold that had to be offloaded. A few flights later, a relatively new rampie discovered a "brick" in the cargo hold and took the heavy object to the supervisor's office. Not knowing what they had, they decided to use the heavy weight to hold a door open. There it sat for years until someone bumped into it and consequently chipped it and discovered its true makeup. Talk about an expensive doorstop!

EVEN MORE UNSUNG HEROES: GATE AGENTS.

It doesn't take long for passengers' stress levels to escalate when a delay is announced. Many will rampage the desk and start interrogating the agent, with questions escalating to demands. And heaven forbid a flight is canceled. I have seen mayhem erupt several times. In recent memory, we had to call the police because one passenger lost it after the flight was canceled due to a huge power outage in Ontario and the American northeast. We couldn't obtain a flight plan. Nor could anyone else.

Sometimes there is only one agent handling a room full of irate passengers. Passengers think if they yell loud enough and threaten with lawsuits — or worse yet, social media revenge — then an airplane will magically appear. I must admit, some agents bring it upon themselves if they are tired, overworked, or not fit for the job, but I've found most people, including agents, want to do the right thing.

SEAT PITCH. IT'S NOT THE ANGLE.

Are seats getting smaller and is the cabin getting more cramped? You bet. But before you start looking to get your money back, here's the scoop. Back in the day, airliners boasted legroom, tons of space for baggage, and phenomenal meals, but airfare was more expensive then. And remember those smoking flights where air filters clogged with nicotine? It wasn't all rosy.

Here are the three parameters by which seats are measured:

1. The pitch on pitch. Pitch is the distance between two successive points. For airline seating, it's the distance between rows, measured in inches; or the space between a point on one seat and the same point on the seat in front of it.

(Pitch is also implemented in talking propeller angle —
equal to the distance forward a blade would move in one
revolution). Generally, pitch ranges from 29 to 32 inches
(74 to 81 cm) on shorter-haul flights, increasing to 33 to 34
inches (84 to 86 cm) on longer flights, but this also depends
on aircraft type. And this affects not just the legroom area;
a bigger pitch number translates into a further stretch of
the upper body. Some low-cost operators have sardine class,
where pitch sits at 28 to 30 inches (71 to 76 cm). I must admit
I didn't know the particulars on pitch, but then again, I have
the best seat in the plane, which is electric, reclines, moves
up and down, and comes with a price tag that could buy a
high-performance car. There are numerous websites out there
comparing airline seating, plus airlines themselves have the
stats too. Pitch is a huge selling component. If you are tall, on
a long flight, or just want more legroom, have the airlines got
a deal for you! You'll get a pitch on the pitch.

2. Then there is seat width — distance between the inside of
the armrests. Something to think about if you have a wider
beam. Economy seats range from 17 to 18 inches (43 to 46
cm), premium buys you up to 20 inches (51 cm), and if you
venture into business class you can squirm around in 21 inches
(53 cm) or more.

3. Recline. This seat angle (no, not pitch) is measured in
inches or sometimes degrees. Recline is from four to
nine inches (10 to 23 cm) in economy. Whether or not to
recline is a bone of contention. There have been major
squabbles over this. And please don't be putting your feet
up on the seat in front of you; your neighbor can feel it.
And be gentle when stowing your tray table, and don't be
pounding on the video screen, because that too can be
felt by the passenger in front of you.

> ✈ During a flight deck visit, one flight attendant
> mentioned that a passenger with very long hair tossed it
> back, covering the video screen of the passenger behind
> her. The poor guy sat there in disbelief, unsure what
> to do next. I firmly believe one could write books and
> psychology theses on human nature and the psyche of
> a passenger. Heck, this could be offered as a university
> course, with flight attendants as the professors:
> psychology 101 at flight level 350.

Seat etiquette goes a long way in maintaining a happy cabin. Yes, seats are getting slightly smaller, but people are getting bigger, wider, and heavier, which is a constant juggling act for airlines. Have you noticed that many shower curtain rods at hotels are now bowed outward? Seats are overall more comfortable, and heck, most now have video screens and power ports, and are much more fire-retardant and offer WiFi.

WHAT IS THE ORIGIN OF "J" FOR BUSINESS CLASS?

Business class evolved in the late '70s. But "B," an appropriate letter for business class, already existed in the computer reservation system, so "C" was next in line. British Airways launched its "club class," but went a step further to "super club," which needed another letter, so "J" came to being for business class. At the time, many airlines used a similar reservation system and adopted "J" to denote "business/ executive class." Although no first class exists with North American carriers, we pilots jest that our vantage points are first class.

Airfare Codes (Airline Class of Service Codes):

A. First Class Discounted

B. Economy/Coach – Usually an upgradable fare to Business
C. Business Class
D. Business Class Discounted
E. Shuttle Service (no reservation allowed) or Economy/Coach Discounted
F. First Class
G. Conditional Reservation
H. Economy/Coach Discounted — Usually an upgradable fare to Business
J. Business Class Premium
K. Economy/Coach Discounted
L. Economy/Coach Discounted
M. Economy/Coach Discounted — Usually an upgradable fare to Business
N. Economy/Coach Discounted
P. First Class Premium
Q. Economy/Coach Discounted
R. First Class Suite or Supersonic (discontinued)
S. Economy/Coach
T. Economy/Coach Discounted
U. Shuttle Service (no reservation needed / seat guaranteed)
V. Economy/Coach Discounted
W. Economy/Coach Premium
X. Economy/Coach Discounted
Y. Economy/Coach
Z. Business Class Discounted

I'M AFRAID OF FLYING. ANY IDEAS TO HELP ME OVERCOME MY FEAR?

You are not alone! Reading books explaining what transpires on an airplane, including those sounds you hear, or taking a fear-of-flying course, goes a long way. Show up early. I see too many passengers in a rush, which only heightens anxiety. And don't watch continuous clips of the show *Mayday*.

> ✈ While boarding on the jetway, I met a passenger that confessed to binge-watching an airline disaster show called *Mayday*. She was "wound up tighter than an eight-day clock." Not a prudent thing to do when shy of flying.

Nearly 30 percent (some claim 40 percent) of the adult population suffers from aerophobia or aviophobia — the fear of flying. It has also been estimated that one in four flyers shows a significant degree of fear or anxiety. And it's not only the less traveled who fall into this group. Many executives and celebrities who rack up thousands of air miles are secretly afraid of flying. Surveys indicate that most fearful flyers do fly, and many are frequent flyers. Why the fuss, since no other form of travel — not even walking — approaches the modern jetliner in safety?

> ✈ Whenever I address an audience regarding the fear of flying, I start off by saying, "Welcome to the second-safest mode of travel!" I then ask the audience what they think the number one safest mode of travel is. I get driving most often, but as mentioned, driving is the least safe mode, especially getting to the airport. The list also includes walking, dogsledding, boating, and biking. But the safest mode of travel is the elevator!

It only takes a few minutes of surfing the internet to realize that an abundance of information and courses are available. There are tapes, videos, newsletters, groups, podcasts, and so on. Some courses have very high success rates. A common tactic is to strategically overcome the fear of flying by knowledge and control, with graduation consisting of a flight on board. Many major airlines offer courses, but some are a tad reluctant to admit there is a problem. Some pilots and flight attendants work on the side offering guidance.

For many passengers, not being in control is a major hurdle. Learning to overcome fear of flying is truly about learning to overcome fear of being a passenger. When we pilots travel as passengers, many ask if we mind sitting in the back, where we are not in control of the plane. To answer that question, ask a flight attendant. They'll tell you pilots are not always ideal passengers.

Knowledge is power. The classroom part of the typical course for fearful flyers is equally divided between aviation education and behavior modification. You will certainly end up knowing more than the average passenger on topics including the theory of flight, meteorology, and turbulence, which is a frequent cause of anxiety. Although turbulence can cause discomfort, it is important to remember it is "generally" not unsafe. An explanation of the aircraft components will enlighten the student on the different noises aircraft make so they can learn to recognize the sound sequences.

Still not convinced? When asked whether we sometimes find flying dangerous, pilots respond, "The most dangerous thing about flying is driving to the airport." My three fender benders in taxis and crew buses make me a firm believer. If you are craving more statistics, well, it's 100 times safer to fly than to drive; you are more likely to die being kicked to death by a donkey than in an air crash; and a fully laden jumbo jet would have to crash every day with no survivors to equal the number of road deaths in a year in the United States.

WHAT HAPPENS WHEN SOMEONE HAS A MEDICAL EMERGENCY ON BOARD?

With hundreds of passengers aboard — a high-density B777 packs in 450 passengers — the probability of someone taking sick escalates. The good news is, all cabin crew have medical training. There is a medical kit that impresses most doctors, including a defibrillator and portable oxygen. And because of large passenger loads, there is a high probability of having a doctor on board. In fact, for all my medical situations, there always seems to be a doctor (or highly trained medical person) on board. Maybe even three or more. Doctors sure travel to a lot of conferences, luckily for the sick passenger and crew. Sometimes they are a tad reluctant to "put their hand up," but the signing of an indemnity form now eliminates the fear of any legal action. However, the decision to divert is made by a company now called Medlink, which offers in-flight medical emergency consultation. I have diverted to a few places under their recommendation, but I've also kept the flight going most times, based on their expert findings. This service is a savior for everyone, because there is nothing easy about "pulling over to the side of the road" in an airliner.

✈ We got the call from the "back end" stating a female passenger had an allergic reaction to possible nuts found in a salad. The flight attendant described her tongue as exorbitantly swollen and looking like a square patty of hamburger meat. Not good. Three doctors emerged, but one overzealous doctor was adamant about an immediate landing. He obviously was following the moving map display and knew St. John's, Newfoundland, was the closest airport. He wanted the patient there, NOW! St. John's is notorious for its inclement weather, and any pilot must be assured good weather prevails before they even think about venturing there. It's one of the meteorological armpits of the world, but a mecca with respect to the friendliest people. Luckily, STAT-MD (our medical consulting company at the time) gave us the go-ahead to Milan, as the patient was receiving as much medical attention as any hospital could give. It all resulted in a happy ending.

HAVE YOU HAD A DEATH ON BOARD?

I realize this is a macabre topic, but in my 35 years of flying as an airline pilot, I have had one death on board. Recently, I met one flight attendant who had two deaths within a month. His co-workers were calling him the Grim Reaper. Many think airplanes are filled with cheerful passengers flying on an annual pilgrimage to all-inclusive Caribbean meccas. But truth be told, many are traveling to attend funerals or seek medical treatment, and some are flying back to their roots to spend their last days. I vacationed on a cruise ship where three people passed away during the week. Some don't realize it, but death and taxes are not optional.

We immediately contact the authorities about a "presumed death on board." Notice I said "presumed." In fact, airports like London Heathrow (just so happens this is the flight on which I had a passenger die) deem that the person has not died until the port authority doctor certifies the death. Only a licensed physician can pronounce death; otherwise it is deemed "apparent."

But what transpires in the cabin? How do flight attendants handle an apparent death? This is yet another part of their job where they go above and beyond, as it is immensely stressful for everyone. If possible, passengers are displaced, but remember, most flights are full. Or, if possible, the body is relocated. The body is covered with a blanket up to the neck, the seat is reclined, eye shades are used, the seat belt is fastened, and pillows are used for padding. This is an extremely sensitive topic, so I will truncate this paragraph and only say that flight attendants do their utmost!

✈ Airlines transport a gamut of goods in the cargo holds, including caskets and human remains. It's a reality of life. One francophone flight attendant told me that during her first year as a new hire, she decided to have lunch on the tarmac in the refreshing sun during a long ground stop. She noticed a long rectangular box with English lettering and sat on it while eating her sandwich. A colleague came along and quickly translated the writing on the box: "Human Remains."

DO PILOTS TAKE NAPS?

Yes, they are called controlled naps or rests and are 40 minutes or less. It's proven that an adequate nap recharges one's batteries. It's a safe and effective mitigation of fatigue. Controlled rest is allowed during the cruise phase and must end 30 minutes prior to descent. It was given the green light a few years ago and is now written into the procedures. Heck, there is even a checklist for it. Can you imagine consulting a checklist before having a nap? We do. It involves telling the flight attendant to call the flight deck at a specific time and to ensure others won't call asking "when are we landing"-type questions. Prior to this, pilots had an unwritten code for telling the other pilot they needed rest before the procedure was written into the scriptures. A pilot may have said, "I am going to look at the above circuit breakers" or "I'm going to look outside for a while."

For long-haul flights, an extra pilot is added, and for those ultra-long-haul flights there are two extra pilots. The long-haul airplanes are equipped with crew rest facilities consisting of two beds and one or two comfy chairs with a video screen. There are even crew rest facilities for the flight attendants in the rear of the aircraft. However, the medium-range airplanes do not have dedicated crew rest facilities, so controlled naps are the next-best thing.

WHAT IS THE MILE-HIGH CLUB? HOW DO I JOIN?

Some may think this is yet another incentive program used to accumulate airline travel points. After all, most airlines offer some form of reward for committed travelers. But ever since the dawn of aviation, people have elevated their promiscuity to greater heights. Because of this, an informal "mile-high club" emerged consisting of people who'd had sex (*ahem*) on board airplanes. I've been on several flights where flight attendants caught passengers in the act or couples sheepishly leaving the lavatory together or moving rather feverishly under a blanket. Usually it involves alcohol, but not always, and sometimes the couple just met. Not sure if it is the excitement of travel, the airplane's subtle vibration, or just wanting to try something different, but the mile-high club exists. There are a few websites where one can tell others about their lofty escapades. Some charter companies also offer membership (signed certificate), if you feel this is a must for your bucket list. But be forewarned, airlines don't have a sense of humor about this membership, and you may be banned from future flights.

✈ I flew with one first officer who had previously flown for an air charter company that, during leaner times, had come up with the idea of offering a mile-high membership in a light twin-engine airplane. They modified the back with a mattress and drew up certificates. He claimed he flew about 30 of these horny missions. On one flight, and after a couple's official indoctrination, they popped their heads through the curtain and asked the pilot whether he would like to join the club. But there were two glitches: one, there was only one pilot, and two, there was no autopilot.

HUMAN TRAFFICKING — MY VERY SERIOUS COMMENTARY.

Every year, aircrew must partake in annual recurrent training, part of which requires pilots and flight attendants to converse for a couple of hours on various topics. Recently, the discussion touched on human trafficking. It is rampant, and I'm still shocked at the magnitude. So much so, I feel compelled to write about it. Mostly young girls are coerced by manipulators and flown to other cities to work in the sex trade. One would think they arrive from other countries, but not always. They may be pressured in Los Angeles and get on a flight to New York. More and more airlines are training flight attendants to recognize the signs. Hotels are also being warned to watch for people being holed up in a room for several days with frequent visitors. McDonald's restaurants and other companies are offering Safe Place, a national youth outreach and prevention program for those under 18 (up to 21 in some communities) in need of immediate safety or help. Flight attendants may only have a brief contact with the potential victim, but you, the passenger, may be sitting next to the victim.

Ask yourself these questions:

Can the passenger speak for themselves, or is the person next to them controlling what they say? Does the passenger avoid eye contact?

Do they appear timid, very quiet, fearful, anxious, tense, depressed, nervous, submissive, or just plain sad? Are they dressed inappropriately? Sometimes their hair is done up and their fingernails manicured, as they may be heading to their first encounter. Do they have carry-on or few possessions, especially on a long flight? Did the passenger go to the washroom alone, or were they dependent on a chaperone to take them to the lavatory? One flight attendant in our class mentioned that a co-worker had noticed an older man holding up a blanket for the girl to change her blouse instead of allowing her to change in the washroom. It turns out the flight attendant's suspicion of "something not right" proved correct. I realize this topic is not a positive side of aviation, but sadly, it's a reality, and it's at epidemic proportions.

SOME PASSENGERS ARE NOT "KUMBAYA" TYPES.

Many envision airline flights to be filled with happy passengers traveling for a fun vacation or just happy to be on an airplane, and most are. But remember, there are all walks of life sitting amongst you. Airlines transport prisoners with one or two accompanying police officers, or deportees with or without supervision. There are passengers who, because of religious reasons, can't sit next to someone of another faith or a different gender, and this can sometimes be confrontational. There are people reeking with body odor or bad breath, people who have not washed in days, barefoot types, clipping or painting their fingernails types, bullies, people watching pornography, and even kleptomaniacs.

✈ One flight attendant said passengers' possessions were going missing during the flight. It started with a pen, followed by an iPhone, culminating with a laptop. Later the culprit was challenged but insisted on their innocence. I've also heard of passengers rummaging through carry-on bags located in the overhead bins. While deadheading, I've seen one passenger walking through J-class on a short flight to New York, asking for religious donations. Most passengers are ideal, but as anywhere in life, there are some that are far from Kumbaya types.

COPING WITH COVID.

As I write this, we are in the thick of COVID. Vaccines are emerging, but most countries are still in lockdown mode and not encouraging travel, producing another nail in the coffin for the travel industry. Already there have been airlines that packed it in, with most teetering on financial ruin. My airline announced another 1700 layoffs.

The world, including aviation, has turned upside down. But flights are still moving. You'll see very senior staff on the ground and in the air as everything in North America is based on seniority. Some of my flights operate with the most junior flight attendant having 25 years of seniority. Over 600 pilots have been laid off out of 4400. Easily another 2000 pilots could be let go, but they have been kept on the payroll for now because of training costs.

Many think an aircraft cabin is conducive to COVID because they believe the air is stale and not circulated. The air in an airliner is exchanged every two and a half to three minutes or about 20 to 24 times an hour. More so than a hospital or an enclosed building. Numerous articles are being pumped out on social media that play up the contagiousness of COVID in airliners. There are also the airlines preaching the opposite, with the facts dispelling the myths. People are reminded of the refresh rate and the HEPA (High-Efficiency Particulate Air) filters that contain almost 99% of airborne pollutants. But many are not buying it.

Airliners are thoroughly cleaned during every stopover, including spraying down the cabin with an electrostatic sprayer. More and more groomers and cleaners meet the flight to scrub the innards of the cabin. I wish I bought stock in plexiglass, masks, vaccines, signage, hand sanitizers, plastic suits, plastic wrap, latex gloves, and cleaners.

You'll see all passengers and crew wearing facemasks. Some anti-maskers refuse to wear one but will be challenged by a senior staff that takes no guff, or if it escalates, they will be deemed a disruptive passenger and/or met after the flight, running the risk of making the "no-fly" list of the airline. I've had several noncompliers on my flights.

You will hear a lot of plastic cellophane unwrapping during meal service as everything is individually sealed. It is disheartening to see the added waste as an environmentalist, but it's for everyone's well-being. For those in economy class, meal service or snacks for purchase have been terminated during shorter flights, with announcements being made in the waiting lounge, "You better go buy something to eat." Coffee, if served, is supplied with no cream but with a powdered substitute. (Actually, milk or cream is available, but you didn't hear that from me.) Wine is served in a disposable cup, taking away the ambience with limited drink options. Passengers once looked at food service as entertainment, but that show is over. If you have the luxury of food service, you have to learn how to do it with a mask close at hand. You'll see flight attendants on long flights twiddling their thumbs as the undertaking of customer service is almost non-existent, and you can be guaranteed a friendly, sometimes curt, reminder your dining is done and you must return to wearing your mask. And let's not forget the barrage of new pre- and inflight announcements related to COVID issues. The lawyers and policymakers are having a heyday writing new scripts to bombard the traveling public with.

The pleasantness of airline travel has succumbed to COVID stress. The fun factor has been nearly depleted by mask requirements, reduced service, and the imposed guilt of traveling. It sure is a dark time in aviation. It makes the past severity of liquids and gels and the removal of shoes into a joke. And what about that orange that would have shut down the country if you didn't declare it? COVID has opened the eyes of many as to what is important and what is superfluous.

Pilots are wearing masks to the flight deck and removing them when the flight deck door is closed. One silver lining is the augmentation of cargo flights. People are ordering online like no tomorrow, making one of the richest men on the planet richer. Plus, many goods must be shipped by air, such as perishables, medicine, mail, and other commodities. These goods must be shipped with a reduced schedule because the world hasn't stopped yet.

My company has capitalized on cargo-only flights by creating a dedicated cargo operation. Even flying with zero passengers, a profit can be made with cargo in the belly on some aircraft. I've flown these nouveau cargo flights, and I must admit it seems weird to walk about an empty cabin of 298 seats. There are new noises and an eerie feeling of passenger apparitions sensed within the empty cabin. Many pilots have voiced the same spookiness. But heck, we pilots have learned how to make coffee, heat our crew meals, and find snacks in the galley with no flight attendants. All is not lost.

THE EUPHORIC SIDE OF AIRLINE FLYING.

Many passengers love the airplane for getting work done at their flying desk. Some authors have written good portions of their books while cruising at flight levels, as flight induces thought. Many of my anecdotes were written while observing as a passenger. Contacts and friendships are made, future spouses have met, ideas have materialized, and numerous songs have been written about flying and its aura. John Denver's "Leaving on a Jet Plane," Frank Sinatra's "Come Fly with Me," "Jet Airliner" by the Steve Miller Band, and my favorite, "Dream No. 2" by Ken Tobias, are some of the many. There is a ton of inspiration oozing about the cabin.

People are moving to better places in their lives, giving talks, meeting relatives and friends, or just seeing the world from a different perspective. It might sound like "Kumbaya," but how exciting!

WILL THERE EVER BE A PILOTLESS FLIGHT DECK?
AFTER ALL, EVERYTHING IN THE FLIGHT DECK IS AUTOMATED.

This question is becoming more prevalent especially with Elon Musk et al. creating autonomous trucks and cars. After all, the airplane should be next. My short answer is: "Not during my career." Sure, there are autonomous trains, but they are one-dimensional travel. Funny, the new train I take to the airport requires two conductors. But the train between the terminals at Toronto Pearson requires none. Why? I guess it's like how Uber and Lyft were first treated by the cab companies. It's new, unique, therefore it must be challenged. I don't think I will see pilotless flight decks, but I do foresee one pilot managing it all. But remember, aviation is three-dimensional travel moving at great speeds, and a lot can make things interesting. For example, in an autonomous flight deck, the weather radar may determine the thunderstorms up ahead are somewhat benign, so the divine computers might fly you through these extremely bumpy clouds. A real pilot would divert around them. On most every flight, there is always something that needs attending to. Now ponder if it was a computer deciding your fate. Would you fly with "Air Autonomous"? Would you be comfortable in saying, "Hey Siri, take me to Chicago," or would you mind hearing, "This is your captain speaking, I am working from home today?" I'm thinking that 30 percent statistic regarding passengers with flight apprehension just got bigger.

✈ They say the future flight deck will have a dog and a pilot. The pilot will be there to feed the dog, and the dog will be there to ensure the pilot touches nothing. Wouldn't it be nice to reduce or get rid of those overpaid pilots? Or as an owner of a now-defunct airline put it, "Pilots are nothing but overpaid, oversexed bus drivers." I could feel the love.

WHAT DOES "KEEP THE BLUE SIDE UP" MEAN?

Blue represents the sky, and brown the ground. The colors meet at the horizon, which is what pilots reference during flight. Blue (upper half) and brown (bottom half) are the two colors for the main instrument in the flight deck, called the artificial horizon or attitude indicator. In fact, it is dead center in the "six pack" of instruments an instrument-rated pilot uses to aviate. This aviation idiom of well-wishing upon departure is akin to "break a leg" for an actor, "keep your stick on the ice" for a hockey player, and "may the force be with you" in the movie *Star Wars*.

✈ CHAPTER 6 ✈

PRE-DESCENT, IN-RANGE, LANDING AND AFTER LANDING CHECKLIST

HOW IN THE HECK DO YOU LAND IN NEXT-TO-ZERO VISIBILITY? I DIDN'T SEE THE RUNWAY UNTIL WE LANDED.

Have you ever looked out of an airplane window as it descends, and you go lower and lower, and wonder when, and if, the ground will appear? Many of us have probably been on flights like this, but just how do pilots find the runway? (Interesting to note, a takeoff requires a higher visibility than a landing.)

A PILOT'S APPROACH.

Despite what seems to be a precarious situation, commercial and some private pilots routinely fly safely into clouds with the aid of instruments. A handful of different instrument approaches are currently available, but the most precise and preferred approach is the ILS (Instrument Landing System), which provides both vertical and horizontal guidance in low-cloud conditions, fog, rain, snow, haze, and other obscuring phenomena.

How does it work? A localizer signal at the far end of the runway guides the pilot or autopilot in a straight line toward the runway, while a glide slope signal on the side of the runway leads the aircraft down vertically. An easy way to visualize this precision approach is to picture a children's slide at a park. The aircraft flies at altitude just as a child sits on top of the slide. The airplane is then eventually steered in the direction of the runway, whereupon the flight deck instruments lock onto both the localizer and glide slope signals. When the aircraft is locked onto both signals, it is as if it is in the crosshairs of a rifle. On board, sophisticated autopilots guide the aircraft all the way to the ground, automatically compensating for changing winds and other variables. The precision approach guides the pilot down to his or her landing site (the runway), just as the slide guides the child to the landing. A localizer provides left-right orientation with the runway, like the sidewalls of the slide. The angle of this approach is typically 3°. It's the angle you may have noticed airplanes maintain while following one another on approach to a busy runway. The glide slope signal guides the aircraft down vertically, and the auto-thrust system adjusts engine power settings to ensure proper speed, even bringing the engines to idle at touchdown.

OTHER IMPORTANT FEATURES OF THE ILS.

Several other components augment the ILS and provide additional safety features for low approaches. These include devices that transmit exact distances from the runway, as well as high-intensity runway and approach lighting (the intensity ranges from a dim setting of one to a power-zapping strength five). Sitting by itself is an RVR (Runway Visual Range) sensor along the edge of the runway. It measures distance seen through obscuring weather phenomena in units of feet, and it gives a very accurate idea of what a pilot can expect to see, or not see.

NOT ALL ILS'S ARE CREATED EQUAL.

There are three different categories of an ILS, differentiated by the

DA (Decision Altitude) and prevailing visibility. DA is the indicated altitude at which a pilot must decide to either continue the approach to a landing or abort it and go around. A Category I ILS has a DH (Decision Height) of 200 feet (61 m) above ground. Most large airports around the world have this type of ILS. DA is determined by a barometric altimeter, which the pilot must adjust to the most recent pressure reading at the airport. Every pilot knows one-tenth of a change in pressure in inches of mercury translates into a discrepancy of 100 feet (30.5 m).

A Category II ILS has a lower Decision Height, 100 feet (30.5 m), and it determines height with a device that bounces signals from the airplane to the ground and back, called a radar altimeter (or radio altimeter). It allows the airplane to descend with a higher safety margin. The last, but certainly not the least, is the Category III approach.

WELCOME TO AUTOLAND.

Category III ILS (autoland) has two levels. The first level brings the aircraft to a mere 50 feet (15.2 m) above the runway, at which time the pilot must make a snap decision. The second fully automated level has no Decision Height, meaning pilots do not look outside and they wait for the bump. It is a procedural necessity: pilots looking outside could cause disorientation. Complete faith is bestowed in the system, which admittedly takes some getting used to. A gamut of requirements must be met to allow such an approach. The ground facilities must have high-intensity runway lights, centerline lighting, various markings on the runway, additional RVR sensors, and backup airport emergency power to ensure the runways and taxiways are lit up and the ILS is functioning, even during power outages. On board the aircraft, sophisticated autopilots bring the aircraft to the ground, automatically correcting for winds all the way to the touchdown. Only major airports have such a system, with most only having the system on one runway. Pilots must be certified to do autolands, requiring checkouts in flight simulators every six to eight months. The airline company and aircraft must also be certified for autolands. As you can see, there are a lot of parameters that must be met, clearly separating the amateurs from the pros.

As well, there are many computers that monitor the aircraft systems to ensure everything is functioning at 100 percent. They even make synthesized altitude call-outs to the pilots.

WAITING FOR THE BUMP.

The absolute minimum visibility for a Category III landing is less than the length of a football field, with next to nothing to see when approaching at speeds of 150 knots (173 mph or 278 km/h). Once air traffic controllers clear the aircraft for a Category III approach, the pilots attentively monitor the automatic systems, overpower the urge to look outside, and patiently wait for the bump. Even with the main landing gear firmly on the runway, the flight deck may still be mired in fog because of the landing angle. From ab initio (initial) training, pilots are taught to trust their instruments; still, autoland requires a much higher level of faith in technology.

Because the system is so accurate, the automatic pilot must be disengaged after landing or else the aircraft will try to reposition itself back on the centerline of the runway. Finding the terminal building in such heavy fog can be a difficult task, but many airports have bright green lights embedded in the taxiways to guide the pilots to the gate, or "follow me" vehicles.

The autoland system truly is a marvel of technology and exemplifies just how technically advanced aircraft and airports have become. Nothing can replace the skill of an experienced pilot, but when extremely poor visibility dictates a Category III autoland, technology rules.

HOW DOES AIR TRAFFIC CONTROL DIFFERENTIATE SINGLE AIRPLANES AND THE NUMEROUS AIRLINES?

Every flight has a unique call sign. For small aircraft, a pilot states their aircraft's registration letters using the phonetic alphabet. For example, NC305B translates to November Charlie Three Zero Five Bravo. Incidentally, any aircraft registered in the United States starts with an "N." In Canada, it begins with a "C." I soloed 42 years ago in an aircraft with the registration C-GNUB (Charlie Golf November Uniform Bravo). For airlines, it can be their name followed by a specific flight number. For example, Delta 125 or United 899. But some airlines have unique names, such as "Speedbird" for British Airways and "Dynasty" for China Airlines. One would assume Air New Zealand pilots to have Kiwi accents or China Airlines pilots to transmit with Mandarin enunciation. But sometimes an airline, perhaps Korean Air, is heard with a Midwest American accent — the sound of an expat pilot.

ARE THERE SPEED LIMITS FOR AIRPLANES?

Yes, there are lots! Below 10,000 feet (3048 m) ASL (Above Sea Level), all airplanes are restricted to 250 knots (288 mph or 463 km/h). A limit of 200 knots is in force when airplanes are within 10 nautical miles (18.5 km) of a controlled airport and when they are less than 3000 feet (914 m) above the ground. Speeds must be adhered to within 10 knots, especially when we are given instructions by air traffic control. The airplane itself also has many inherent speed limitations for flaps and landing gear extension, and tire speeds, to name a few. Even windshield wipers can't be operated when the aircraft speed is above 230 knots.

HOW DO AIRPLANES CALCULATE THEIR SPEED?

Speed relative to the air is calculated by instrument probes and sensors on the exterior of the aircraft, near the nose. That's those "whiskers" you see near the front. These pitot tubes face the oncoming airflow (ram pressure), and static pressure is subtracted to determine indicated airspeed. Computers then calculate true airspeed by factoring in such variables as temperature and pressure. Airliners also calculate Mach speed by dividing true airspeed by the speed of sound. Typical Mach cruise speed for an Airbus A320 is .76 to .80, which is equivalent to 76 to 80 percent of the speed of sound. My B787 zips along at a speed of Mach .86. It's fast!

HOW ARE THE OUTSIDES OF PLANES CLEANED?

Some airlines do not have a dedicated outdoor cleaning facility due to climate constraints — aircraft can't be washed when outdoor temperatures are below 4°C (39°F). Because of this, every 18 to 24 months, during heavy maintenance checks, or when an aircraft is deemed in need of a wash, it's pulled into the hangar and washed from nose to tail. Soap and degreaser are applied using sponges and soft-bristled brushes, while pressurized water hoses are used to thoroughly rinse the aircraft. To this day, I would bestow the "cleanest fleet award" to Lufthansa. Though they have a modest livery, their exteriors sparkle.

HOW DO PILOTS NAVIGATE THE RUNWAY DURING THE DAY AND NIGHT?

Instrument Landing Systems and on-board instruments that assist with visual navigation are used for both nighttime and daytime landings. Most runways have special lights off to the side and near touchdown that guide us vertically — these lights are designed to change from red to white depending on the angle to indicate the slope of the aircraft's approach to landing. One version is the PAPI (Precision Approach Path Indicator). There are four lights in a row off to the side of the runway. If the two inner lights are red and the two outer are white, the airplane's profile is spot on. If all four lights are white — too high. If all red — too low. Another version is the VASIS (Visual Approach Slope Indicator System), which works on the same principle as far as color and angle, but two lights are positioned behind the first two lights. No matter the system, the adage goes: "All white you are too high, red over white you are all right, and all red you are dead."

DO AIRLINE PILOTS FLY UNDER INSTRUMENT FLIGHT RULES (IFR), OR DO THEY SOMETIMES FLY UNDER VISUAL FLIGHT RULES (VFR)?

All airlines except the smaller local ones, such as those that island-hop in the Caribbean, conduct all their flights under IFR. Flight dispatch creates and files flight plans with air traffic control. This way, flights are monitored from pushback to gate arrival. While we can occasionally maneuver visually while on approach to land, we don't depart and fly VFR like small airplanes. Having said that, it doesn't stop us from enjoying the view!

WHY DO THE INTERIORS OF ALL AIRPLANES IN AN AIRLINE LOOK THE SAME, REGARDLESS OF MANUFACTURE DATE?

When an aircraft is purchased, specs and customized materials are sent to the manufacturer to ensure that elements such as the distinctive interiors and entertainment system are incorporated to standard. For older aircraft, refurbishment takes place during major maintenance checks. It can take 30 to 60 days to install new seats, galleys, washrooms, closets, and interior walls. Partial redesigns to the overseas fleet — such as the installation of lie-flat beds — require 15 to 30 days and are executed during "special visits."

WHY ALL THE FUSS ABOUT CELL PHONES AND OTHER TRANSMITTING DEVICES? DO THEY REALLY INTERFERE WITH THE AIRPLANE?

Who knows? That's my quick answer when asked about whether cell phones or the like will cause the aircraft to veer off course or will interfere with our instrument landing system. For such a small entity, they sure attract lots of attention. My feeling is technology has advanced so fast, we haven't caught up with what to say and what to enforce. In my first book, I mentioned that airlines intended to replace the "no-smoking" sign with the "okay to use the cellphone" sign. That hasn't really happened. Maybe airlines are reluctant to accept the use of cell phones because of the social implications. Think about it: you just allowed unlimited cell phone usage, which allows anyone to talk impetuously on the phone. Some passengers don't want to hear boisterously loud phone conversations about how many times your dog defecated that day, or listen to those with hearing issues talk about their grandkids, or whether they turned off the iron upon leaving for the flight. People want peace and silence — not loudness, jabber, and bombastic tripe.

If you delve further into cell phone use, you will find that usage at altitude jams relay stations. It turns out the ban on wireless devices has a lot more to do with possible interference for ground networks, rather than any danger posed to aircraft systems. As well, cell phones used to be the size of bricks and transmitted with more power.

On your next flight, notice how many phones become active on descent, receiving texts for those naughty noncompliant passengers who did not set to airplane mode. It's a tough one to govern.

✈ On a very recent flight I took as a passenger, the incharge flight attendant took me aside and divulged the true reason why our flight was over an hour late. Most airlines have come onside about the perils of passengers retrieving their cell phones when dropped into the innards of an airplane seat (it could cause a fire), particularly the large business class seat. A passenger (some airlines use "client") dropped his cell phone. It turns out it fell not into the seat but under the forward foot area elevated nearly an inch high above the floor. The sleek cell phone then slid several seats forward upon landing. After 45 minutes of searching, someone decided to call the phone. It turned out the elusive phone had landed three rows forward. This cost the airline over an hour's delay and thousands of dollars in repercussions, with several passengers fretting about connecting flights. Murphy's Law raises its ugly head occasionally in the airline industry.

RAINDROPS KEEP FALLING ON MY HEAD?

Some passengers quietly sitting in their seat suddenly experience mystical water droplets dripping on them. Is this from a leaky water pipe? No, it's condensation. The aircraft's skin gets mighty cold — remember, the average outside cruising temperature is –57°C (–71°F). And when humidity, mostly from passengers' bodies, contacts the cold aircraft skin, frost or ice may first form. But during descent, the frost may melt. The interior does have an insulated covering, but sometimes water has a way of funneling between the interior panels and out onto an unsuspecting passenger. And yes, this wet surprise can happen in the flight deck too.

ABSURD QUESTIONS.

Some people take weird joy in asking absurd aviation questions. And they are not kids. I just smile and try to grin and bear them. Some of the absurdity: Can my airliner do a barrel roll? Can I fly inverted like Denzel Washington in the 2012 movie *Flight*, which portrayed the airline pilot to extreme absurdity? What would I do if both my engines fell off? Do I have mile-high memberships with flight attendants? Could I fly under the Golden Gate Bridge? Can someone jump out of an airliner with a parachute? Why don't airplanes have parachutes? Have I seen UFOs? Does my airplane spew chemtrails? Can turbulence knock a wing off?

But I will answer this one: Can doors be opened in flight? The pressure difference between the outside and inside is about 8 to 9.5 PSI. A typical cabin door, acting like a plug in a sink and measuring three by seven feet, experiences over 24,000 pounds of force pushing it closed. So, no, one would have to be Hercules and then some.

WHAT'S MORE FUN TO FLY, A LARGE AIRCRAFT OR A SMALL ONE LIKE A CESSNA?

As an airline pilot, it's reassuring to know that a bunch of departments work behind the scenes to ensure everything goes smoothly. When I flew small single-engine airplanes, I couldn't fly into cloud and had to navigate referencing the ground the entire time, with little instrumentation. I also had to look after everything from filing my own flight plan to fueling the airplane. But, then again, flying close to the ground gives you vantage points second to none! I have tried twice in recent years to qualify to fly small airplanes, but both times I left feeling apprehensive — a common trait amongst airlines pilots. The irony is, I taught people to fly in these small planes years ago.

DO YOU TALK BY RADIO WITH PILOTS FLYING THE SAME ROUTE AHEAD OF YOU TO GET INFORMATION ON WEATHER CONDITIONS, TURBULENCE, ETC.?

When flying the busy air routes, we communicate solely with air traffic control and relay flight information through them. We also monitor an international emergency frequency on our second radio. Over sparsely settled areas, such as the Arctic or the Atlantic or Pacific Ocean, pilots communicate to each other on a specific frequency. We pass on flight and weather conditions and, occasionally, the latest sports scores on a frequency easy to remember: 123.45 MHz.

IF YOU WERE ASKED TO BUILD OR RUN AN AIRPORT, WHAT WOULD YOU HAVE ON YOUR WISH LIST?

- A fast, efficient transit system to the airport is near the top of the list. While waiting for the crew bus in LAX, I am shocked by the number of cars, which turn the roads in and out of the airport into parking lots. A huge percentage of those cars are Uber and Lyft, meaning a good proportion of people in LAX are moonlighting. LAX is building a new system that will at least move the car rental buses to an offsite location with an automated people mover. Pity, I will be retired by the time it is completed. As I write, LAX now has a new pickup area for Uber and Lyft. You must take a free shuttle bus, but I think that change helps. That's what I love about my present residence; I get to take the UP Express (high-speed train) from downtown Toronto to the airport. It's the best! Many European airports are with the times: London Heathrow, Charles de Gaulle, Hong Kong, and many others offer fantastic bus and train services.

- Gates that allow airplanes to be pushed back without interfering with other gates and airplanes. All it takes is one aircraft to move into a tight alley and go mechanical (break down), preventing others from moving. Many airports around the world are plagued with this hindrance.
- Free WiFi. I can't believe how many airports want to nickel-and-dime the passengers by charging for this service. Don't they make enough from landing and airport improvement fees? And don't offer the free service following a questionnaire or require the passenger to provide the server with an email address so the airport can hound them later. (My son suggests making up an address. As long as it contains an @ symbol, it should satisfy their yearning.)
- Comfortable chairs. I am shocked by how many airports have those uncomfortable chairs with armrests welded into an unwelcome position. Yes, I realize airports don't want passengers to think of an airport as a hostel and prep for an overnight cheap stay, but at least offer a few seats with lie-down options. And offer more seats! Not with iPads stuck to them trying to sell passengers outrageously priced junk food.
- Lots of washrooms. All it takes is for a few toilets or urinals to go mechanical, together with delays due to weather or air traffic control, and you'll realize the importance of such amenities. Especially if, for some crazy reason, you are compelled to sleep overnight at the airport.
- Healthy food options — not sugar-coated donuts and tar-like coffee. And keep the prices down. I know many passengers, including me, would never venture into an airport restaurant, knowing full well the prices are elevated astronomically to offset the ridiculously high rent.
- Lots of power hook-ups. How many airports have I seen that are in denial about cell phone and computer usage?

Many make passengers stand against the wall; however, some passengers plunk themselves down on the floor, causing traffic congestion.

- A welcoming smile at an information kiosk. Sure, the airport is plastered with signs, but it sure is nice to talk to a human about taxis, available hotels, the nearest washroom, etc.
- Lower the noise. Get rid of the TVs blaring 24 hours of depressing news from CNN or Fox News. Maybe get back to airport music? I've been hearing pleasant music emanating in some sections of Toronto Pearson.
- Allow passengers a view of the airplanes and airport. An airport is dynamic and full of upbeat people and energy. And put windows in those claustrophobic, closet-like jetways (bridges). What better way to kill the joy of travel than by sending passengers down a rectangular tube akin to a series of rusted railroad boxcars welded together? I know, I know . . . wishful thinking.

My list is far longer than this. Most airports should consult airline staff to improve conditions. I can readily name some airports where most of the staff, who are there for the so-called well-being of passengers, should be let go. They just don't get it.

HOW OFTEN ARE AIRCRAFT REPAINTED, AND WHAT ACCOUNTS FOR THE DIFFERENT COLOR SCHEMES?

Generally, aircraft are repainted every five to six years or when the livery is changed to jazz up the look. A handful of aircraft may be chosen to support special events or initiatives. If an airline is selected for a global event such as World Cup soccer, you can rest assured they will be sporting a new look. Alaska Airlines sure have unique brightly colored tails, with many depicting pictures of animals from that state.

DO AIRPLANES HAVE ABS (ANTI-LOCK BRAKING SYSTEMS) LIKE CARS?

Airplanes are equipped with an anti-lock braking system, but we call it an anti-skid system.

In fact, it was first invented for airplanes. Even motorcycles now have ABS to keep tires from skidding on both dry and slippery surfaces. The anti-skid system is incorporated into the brake system on the main landing gear. The nose wheel has no brakes but steers the airplane on the ground.

RADIO WAVES.

Pilots converse via two VHF (Very High Frequency) radios. One radio is dedicated to air traffic control whereas the second is for company operations, and during cruise, we monitor an emergency frequency. VHF is limited in distance due to the curvature of the Earth, so for long distances, HF (High Frequency) radios bounce airwaves off the ionosphere. The wide-body fleet is equipped with satellite communication. During taxi in London Heathrow, I've talked to maintenance in Montreal, and while en route to São Paulo consulted a doctor sitting in Pittsburgh about a medical issue. WiFi is becoming more and more available in the cabin and is now available in many flight decks.

Chatty facts:

- VHF is restricted to line of sight. At high cruising altitudes, this equates to about 250 miles (400 km).
- "Five by Five" is pilot talk for excellent radio reception and readability. Strength and readability are ranked on a one-to-five scale. I knew one pilot who had "5 by 5" as a personal license plate.

- Pilots must have an ROC (Restricted Operator Certificate) to communicate on aircraft radios. I also have a marine radio license.
- Aviation uses the international phonetic alphabet, so every letter in the alphabet is represented by a word. DOUG is spoken as Delta Oscar Uniform Golf. Pilots and air traffic controllers know these terms verbatim. When teaching brand new pilots years ago, I told them to practice the phonetic alphabet by saying the letters found on car license plates.

A	Alpha	N	November
B	Bravo	O	Oscar
C	Charlie	P	Papa
D	Delta	Q	Quebec
E	Echo	R	Romeo
F	Foxtrot	S	Sierra
G	Golf	T	Tango
H	Hotel	U	Uniform
I	India	V	Victor
J	Juliet	W	Whiskey
K	Kilo	X	X-ray
L	Lima	Y	Yankee
M	Mike	Z	Zulu

- The word "heavy" is appended to our transmissions on wide-body aircraft, denoting that the aircraft is in the 300,000-pound-and-over category. Some of the more popular aircraft types labelled "heavy": Boeing B747, B767, B777, and B787, as well as the Airbus A330, A340, and A350. The no-longer-manufactured B757 is deemed "heavy" because of its notorious wing tip wake but does not actually fit into the weight category.

- Another category — "Super," for aircraft over 1.2 million pounds (560,000 kg) — is used for the goliath Airbus A380. There is also a six-engine Russian-made heavy-lift Antonov in that weight class, but it's not an airliner.
- NORDO is the acronym for "No Radio." Some small privately owned aircraft don't have radios and can only fly in limited airspace.
- If all else fails, the tower has a light gun that can be aimed at the aircraft to communicate. A steady green light means "cleared to land!"
- To fly a flight from Toronto to Montreal or LaGuardia to Logan, Boston, requires switching frequencies at least 15 times on VHF radio.

JETTISONING FUEL. WHY WOULD A PILOT BE GETTING RID OF FUEL?

Large aircraft (wide-body — the ones with double aisles), which tend to go long distances, require lots of fuel. So much so, if the pilots must land prematurely, perhaps for a mechanical reason or medical diversion, then the aircraft may be too heavy. To alleviate this, fuel can be jettisoned overboard at about a ton per minute to lighten the aircraft. This "fuel dumping" procedure is done in designated areas and at certain heights (above 5000 feet, or 1524 m) to disperse the fuel before it settles to the ground. The fuel is jettisoned from the main tanks from a pipe on the outer portion of the main wing. The narrow-body fleet (one aisle) do not have this option.

HOW HECTIC IS IT FLYING INTO NEW YORK,
WITH THREE SUPER-BUSY AIRPORTS CLOSE BY?

Not only are there three extremely busy airports — EWR (Newark), LGA (LaGuardia), and JFK (John F. Kennedy); four if you include

TEB (Teterboro, New Jersey) — within a short distance of each other, but the northeast quadrant of the United States is the busiest airspace on the planet. My hat goes off to the pilots who must converse in English as their second language, especially for those concluding a long overseas flight. Many New York and Boston air traffic controllers have unique accents or rush their transmissions, and sometimes the controller's patience is tried. But they are moving billions of dollars of aluminum. I've heard pilots being reamed out by them. Luckily, my flying now excludes this intense, densely packed jabber. I only skim Boston's airspace while setting course to Europe. But Europe has the second-busiest airspace, and what makes things interesting there is that much of the air traffic control appends an extra one or two digits to the radio frequency. For example, ground control could be 121.9 or 121.65 MHz in North America, but in Europe it may be 122.705. It doesn't seem like much, but you'd be amazed how often the frequency exchange is repeated because of the extra digit.

Flying through this airspace requires frequently changing radio frequencies as one controller passes you off to the next — usually to the guy sitting next to them.

I liken the ending of a flight to nearing a vortex. At first, radio chatter is casual, switching from controller to controller as you traverse the landscape, but as you near the airport to land, frequency changes speed up, chatter is more frequent and louder, and you better be listening.

WHY DO AIRPLANES NEED TO TAKE SUCH LONG, GRADUAL APPROACHES/DESCENTS?

Generally, the rule of thumb as to when to descend is: multiply the altitude by three, with the number equating to distance. So 30,000 feet (9,144 m) times three means we should start down 90 NM (167 km) away, in a no-wind scenario. Luckily, computers on board do the math, but I still find myself doing the 3:1 ratio on descent. We also must adhere to speed limits and altitude and noise abatement constraints. Plus, we must allow time to depressurize — also done automatically on airliners.

THE FINAL APPROACH FIX (FAF).

This is the final navigation waypoint a pilot passes before landing. They range from three to six miles (5–11 km) back from touchdown. Approaching and passing this crucial waypoint is a very busy time in the flight deck. Nearing this point, the pilot is slowing down, extending the flaps and landing gear — configuring the airplane for landing. There is constant chatter — both with the approach controller, who is ensuring the arriving aircraft fit in the lineup, and within the flight deck. Right over the checkpoint, the altimeter reading is checked and the missed approach altitude is confirmed set. Around this point, a radio frequency change is made to converse with the control tower — the place where most people wrongly envision the entire air traffic control system to be nestled. Most airlines have stable approach criteria consisting of two steps. For the first step, the aircraft better be configured to land. The second step requires airspeed within a certain range, proper height above the ground, little lateral deviation, and a rate of descent that isn't too fast. If these criteria aren't met, a go-around must be initiated. Near this busy fix (waypoint), the pilot receives clearance to land, which is music to their ears. Because this is a climactic point, you'll hear pilots reconfirming their clearance to land even though they were indeed cleared to land. When in doubt, ask!

WHY DOES THE LANDING GEAR EXTEND SO EARLY?

The timing varies. Sometimes it takes over five to six minutes from gear extension to touchdown because of aircraft type, speed, weather conditions, etc., but generally landing gear is extended about two NM (3.7 km) prior to the final approach fix, where we should be configured to land. This varies from airport to airport and runway to runway. For example, landing in Narita (Tokyo), we must extend the gear before crossing the shoreline, just to alleviate the threat of ice falling from the landing gear — which means that about 12 NM (22 km) out, the gear will be dangling. And yes, there are people monitoring the procedure.

LANDING VOWS.

Prior to descent, the "pilot flying" must brief the "pilot monitoring" on the arrival and approach, runway details, flap and brake settings, weather considerations, etc. The list is extensive, and everything is recorded. I refer to these briefings as our "aviation vows." Winds and traffic volume are two major players in runway designation, and air traffic control can ask a pilot to switch runways at the last minute, especially at major airports.

Some light trivia:

- When teaching landings to ab initio student pilots, an instructor looks for three things: landing on the main landing gear first, on the first third of the runway, and on the center of the runway. The rest is gravy. This also applies to landing the big ones.
- To efficiently cater to arrivals, a STAR (Standard Terminal Arrival Route) is implemented. Many of the approach names are themed, such as the Goofy, CWRLD (Sea World), and PIGLT (Piglet) arrivals into Orlando, Florida. Ottawa has the Capital and Phoenix has a DSERT (desert) arrival.
- "Cleared to land" are words all pilots must hear, and a smooth landing is colloquially known as a "greaser" or a "smoothie." Avoiding a thud on landing is easier on a wet landing strip.

HOW DO THRUST REVERSERS WORK TO SLOW DOWN AN AIRCRAFT?

Thrust reversers are mechanisms that reverse the airflow to slow down the airplane after landing. Traditional "clam shells," or "buckets," located where the exhaust exits, have given way to newer technology: modern aircraft have "petal" reversers — so called because they take the shape of a flower when extended outwards — which reverse airflow along the sides of the jet engine.

WHAT IS A GO-AROUND AND WHY THE STEEP MANEUVER?

Major airports are super-busy. Airplanes are spaced as close together as safety will allow, which equates to three to five miles. Because of it, many airports operate in a HIRO (High Intensity Runway Operation) mode. Translation: move it to the runway, expedite the takeoff. For landing, get off as quickly as safely possible. There is no time to finesse a smooth landing, but we pilots will try our best given the limitations. Sometimes the preceding aircraft may have exited the runway a little slower than expected (trying to grease it on), so a safe landing is not possible. We therefore must execute a go-around or missed approach. The aircraft angle is steep, with full power rumbling from the engines. A go-around may also occur because the pilots were just not ready because the aircraft was not configured to land. Weather also plays into it. Tailwinds aloft are a huge factor, because they make it more difficult to slow down. There is a saying when flying jets: "You can't go down and slow down," especially with sleek modern airliners. I admit a go-around is abrupt, but it is very safe. True, it can be a shocker — here you are, gathering your thoughts about what to do after the flight, and suddenly you're wondering what the issue is as the nose points upwards. Think what the pilots are thinking and how their workload just increased exponentially.

WHAT'S WITH ONE MAIN WHEEL TOUCHING DOWN BEFORE THE MAIN WHEEL ON THE OTHER SIDE? IS THIS A POOR LANDING?

As mentioned, pilots prefer to land into the wind, but frequently winds blow across the runway. To keep the aircraft aligned with the centerline of the runway, a pilot must maneuver the flight controls in a way that keeps the main landing gear oriented straight down the runway, so the main landing gear upwind will touch first. This is a normal procedure and is much safer than if the pilot did not correct for the crosswind. Pilots are taught, "Turn into the wind and use the opposite rudder." This technique works for anything from a Cessna 172 to a Boeing 777. There are some modifications to this, but I keep the procedure simple. Truth be told, a pilot is more likely to really ace the landing with a slight crosswind compared to calm winds. It takes finesse to perform a crosswind landing, especially when the winds are strong, gusty, or blowing significantly across the runway. Every aircraft has its own inherent crosswind limitations, so sometimes a landing can't safely be performed. It's either use another runway or go to another airport.

✈ Years ago, while on final approach to a gusty crosswind landing, the captain looked over to me and asked how my crosswind landings were, as I was new to the company and airplane. I told him curtly, "You are about to find out." He smiled.

AIRCRAFT ATTITUDES.

During initial takeoff, we rotate to about 10° to 15°. When we clean up the aircraft, i.e., retract the landing gear, flaps, and leading-edge slats, we climb away at about 5° above the horizon, slowly decreasing the attitude as we climb higher. At cruising altitude, both the Airbus and Boeing cruise at a 2.5° above the horizon. (Who says these two rivals have nothing in common?) During descent, it's only about 0° in relation to the horizon, and sometimes a couple degrees below. On approach, the flight path angle to the runway is a standard 3°, but our nose is near the horizon.

LANDING THE BIG ONE.

Landings are like golf shots. Sometimes you wish you could redo it. Yes, I've been there. Everything is aligned for a nice landing, and then bam! It bites you right in the ass. It is embarrassing, and every pilot has been there. If you find a pilot who states otherwise, they're lying. About a decade ago, I had a snug (firm) landing. No, the oxygen masks did not fall, nor did the overhead bins open, it was just snug. One passenger decided to capitalize on the situation and wanted to file a report. The gate agents got involved, so I went up to the gate. The passenger claimed his arm was hurt during landing. Hmmm? I could tell he was fishing. A few minutes later, he confessed to losing his entire pension through bad financial decisions, so it seemed he was using the situation to sue. I curtly told him the landing was not that hard, I've seen worse, and he quickly slithered off.

WHAT IS A HARD LANDING?

During a recent interview for a magazine, I was asked to explain a hard landing. Why is it that at some airports, passengers know to expect a "thumper"? LaGuardia comes to mind in the snug landing department. I explained that the word "hard" should be replaced by "snug" or "firm." Hard would mean "planting it on" and would require a maintenance check, which includes looking at the "G" value at landing.

Most airliners have that ability. I used to teach Airbus pilots the ropes, and a few landings neared that threshold.

There are several reasons why you've encountered hard landings. As mentioned, many airports operate at HIRO (High Intensity Runway Operations) capacity, meaning "get on, get off." There is no time to finesse a greaser. If you do, you may cause an airplane to go around, or you might piss off ATC because operations are tight. If the runways are contaminated with snow, ice, or significant water — called standing water — then a pilot wants to aim for a point, get the airplane on to activate the ground spoilers, reverse thrust, and apply the brakes — though some may over-emphasize this maneuver. Aviation is on the rebound and growing exponentially, especially in the U.S. Thus there will be new pilots behind the wheel. And where do many get sent to learn the ropes? Yup, LaGuardia or an equivalent. LGA not only has shorter runways, with water staring at you on three out of the four runway ends, but the two runways intersect, making the operation even more difficult.

Carrier landings are up there in the hard landing category. Naval and marine aircraft carrier pilots regularly touch down at 800 feet (244 m) per minute to hook the arrestor cable, compared to a much softer 50-to-150-foot-per-minute airliner landing. One major American airline has had enough and is now denying naval and marine applicants. Because of their training as carrier pilots, they are more likely to land a thumper than a greaser.

I could chew up a page making excuses for less-than-ideal landings. But thankfully most landings are greasers, smoothies, "have we landed?" or "a beauty" (as in that was a beauty!). Yes, a smooth landing strokes a pilot's ego immensely. It's why we land the plane manually 99.5 percent of the time. We mostly save the autoland for low visibility. You will see the pilot saying goodbye to passengers when the landing was acceptable to good, but many will remain in the flight deck if it was otherwise. Yup, been there as well.

Step-by-step landing:

- Follow instructions from air traffic control to fit in with other landing aircraft.
- When cleared for the approach, arm the approach button to guide the airplane vertically and horizontally down to touchdown point.
- Slow aircraft by selecting desired speed, and about 10 NM (18.5 km) out, request first flap setting.
- Prior to the final approach fix, ask for the landing gear to be selected down (i.e. lowered) and ask for more flap.
- Keep autopilot and auto-thrust/auto-throttle engaged.
- Ensure landing clearance from control tower by hearing "cleared to land."
- Respond by saying "roger" when airplane's digitized voice tells you "100 feet above."
- Airplane calls "minimums," and you respond by declaring "continue."
- Disengage autopilot.
- At 50 feet (15.24 m) above ground, an Airbus airplane tells you to retard the thrust levers, whereas on the Boeing you slowly reduce the thrust.
- Simultaneously begin to arrest the descent.
- Hold it off and wait for the bump.
- Allow nose gear to touch lightly.
- Engage thrust reversers; allow auto-brake to slow airplane.
- Disengage auto-brake, stow the thrust reversers, begin turning off runway with hand-controlled tiller.
- Wait for the other pilot to say, "Nice landing."

VACATING THE RUNWAY — THE AFTER LANDING CHECKLIST.

Upon landing, we must vacate the runway expeditiously at busy airports. There is no time to dilly-dally. This quick-paced transition from flight to ground — switching radio frequency from tower to the ground controller, and control from pilot flying / pilot monitoring to captain and first officer — happens within seconds, and it can be hectic. I tell the first officer when taking control, "We are on taxiway so-and-so and we are going to gate such-and-such." It helps them get their bearings, because they are the ones that talk on the radios while on the ground. And I make sure the next appropriate frequency is in the backup selection. Now is not the time to select the wrong frequency, especially if we must hold short of a runway or we are cleared across an active runway. A taxi incursion, or worse yet a runway incursion, is a serious issue. It's another one of those things a pilot has nightmares about. Sometimes I don't think ATC realizes the busyness (workload) during this transition, but I understand they have a job to move airplanes. Because of it, we must listen up! When clear of the runway, the captain will turn off those blinding landing lights and initiate an after landing checklist performed by the first officer. Flaps are selected up, the spoilers are stowed back into the wing, the APU is fired up, and switches are returned to their correct position. Thereafter, the blood pressure and heartbeat subside, and soon the parking brake will be set to finalize the flight. Then, "It's Miller time!"

✈ CHAPTER 7 ✈

POSTFLIGHT CHECKLIST
AND REFLECTIONS

AIRPLANE DOCKING — VDGS (VISUAL DOCKING GUIDANCE SYSTEM).

Pinpoint accuracy is required to park an airliner. We are either guided to the gate by a ground marshaller using fluorescent (sometimes lit) wands or by an automatic guidance system at larger airports. The VDGS (Visual Docking Guidance System) directs us to within inches of the mark, by signaling to steer straight, left, or right, or to slow down. This electronic guidance varies by country and even airport to airport. Newer VDGSs use a video sensor that detects the aircraft with high-capacity imaging and compares it to the system's 3D aircraft model database. Captains maneuver the airplane by a hand-operated tiller that pivots the nose wheel. Next time you're at the gate, look outside to see if the pilot nailed the center of the yellow line.

✈ After setting the parking brake at the end of a flight inbound from Milan, I get a call from a middle manager the next day asking if I missed my mark by two feet. Someone was pointing fingers, and I wasn't biting. The manager did admit there were several reported overshoots. After keeping it smooth through rough air over the Swiss Alps, dodging thunderstorms over France, and navigating the tracks over the Atlantic Ocean, I get called regarding the last few inches of my nine-hour, 4100-mile flight. You never get called when you do a great job, but that's show biz!

Docking facts:

- The taxi to the gate is under our own power, but when leaving the gate, we must be pushed back by a tractor or tug.
- Some turboprop aircraft can back up by changing the angle of the propellers, but this maneuver is rarely used.
- The VDGS is so advanced at some airports, it allows us to maneuver the aircraft to the gate during a thunderstorm advisory, when ground handlers are absent for safety reasons.
- At one time, a few airports implemented a large mirror at the gate, allowing the captain to manipulate the airplane by looking at the nose wheel reflected in it. But lately the mirror has given way to the VDGS.
- The B777 has a camera system that allows the captain to fine-tune nose wheel placement during ground maneuvers. The A380 and the B747-8 have it too.
- If the VDGS fails or is absent, we rely on a trained marshaller using fluorescent wands to guide us. Some of the American marshallers manipulate their wands as if they are fresh off launching jet fighters from an aircraft carrier. It's neat to see.
- The auto self-parking (VDGS) would be a great thing to have for some drivers when parking their car, or for me when I dock my boat. Maybe I wouldn't have hit my neighbor docked beside me.

- At major airports in Canada and the U.S., the ground marshaller guides us initially and then hands us off to the electronic guidance system. But more and more airports abroad rely entirely on the VDGS, no marshaller required!
- One system, called Safedock, guides aircraft to an accuracy of four inches (10 cm) using invisible infrared lasers to acquire the aircraft's position and type for easier docking.
- Once at the gate, the parking brake is set and the engines are shut down. The captain gives hand signals to the lead ground attendant to confirm the brakes are set and the engines are off. Meanwhile, wheel chocks are set in place, the jetway (bridge) approaches the cabin door, and external power is plugged in.
- When leaving the aircraft, I always peek back to see how close I got to the yellow centerline. For the ego, nailing the yellow line is right up there with a smooth landing.

WHAT'S WITH THOSE JERKY, GANGLY, SOMETIMES UNRELIABLE JETWAYS (BRIDGES)?

If you travel a lot, you will inevitably encounter a jetway deciding it doesn't want to move or doesn't want to move the way it is told. You arrive at your destination, and you want off — fast! You've had a great flight, only to be told, "The airport staff is having difficulty with the jetway." You want to blame the airline for such incompetency, but it's out of the control of the airline, as most airport gates and jetways are controlled and run by the local airport authority.

IF YOU COULD FLY ANYWHERE TO GET YOUR FAVORITE FOOD, WHERE WOULD YOU GO?

The list is long: Germany for Schweinshaxe (pork knuckles); London for Thai food, although I'd also indulge in the "pub grub"; Tel Aviv for hummus; Japan for gyoza and miso soup. I love my chicken tikka masala (chunks of chicken marinated in spices from room service in New Delhi), the gastronomic diversity offered in Hong Kong, plus Seoul's Korean barbecue. Domestically, I'd set out for east coast halibut and heavily laden garlic chicken at the Stinking Rose in San Francisco.

WHAT IS THE MOST UNUSUAL FOOD YOU'VE EVER HAD, AND WHERE DID YOU HAVE IT?

I've had octopus in Greece, steak tartar in France, and chewy cod tongues in Newfoundland. I couldn't get up the nerve to eat freshly skinned eels in Japan or chocolate-coated crickets in Korea. And while I passed on the "thousand-year egg" (preserved chicken, duck, or quail egg) and the "five flavors of dog" in China, I did give pigeon a try in Hong Kong.

TIME TO REST.

Many people, including my wife, think because aircrew fly all the time we don't get jet lag. As one senior captain said, "You always succumb to jet lag, but you get accustomed to it." A co-worker of mine claimed his neighbor would always challenge him to a game of squash the day after he arrived from overseas. He always lost. My motto is never operate anything dangerous like saws or lawnmowers after a long flight. Jet lag affects everyone differently, and knowing what works for you is the best remedy. For me, it's exercise. I've worked out in gyms all over the world: India, China, Korea, England. I remember working out in Japan immediately after a 14-hour flight, but that was pushing things; my lights were on, but no one was home. If the gym sounds too energetic for you, just going for a walk will do wonders to clear your foggy brain. It's what most aircrew do, with many including mega-shopping binges in their walks. Yet another benefit of working for an airline — international shopping.

> ✈ After a 14-hour flight, we four pilots decided to
> visit the markets in Shanghai, as pilots and flight
> attendants love deals. (Or was it Beijing? Jet lag
> induces amnesia too.) Almost immediately, one of the
> pilots inadvertently tried to walk through a glass door.
> His nose took the brunt of it, with blood adding to
> the drama. Not sure what the passengers thought the
> next day seeing him bandaged, bruised, and reddened
> with embarrassment.

THE PRO'S GUIDE TO JET LAG.

Some know the symptoms, but others may not recognize the effects of jet travel: lethargy, dehydration, disturbed sleep (which includes short deep sleep followed by restlessness), moodiness, irritability, trouble concentrating (trying to convert currency in your head is a big one), sluggishness, possible headaches, dry mouth, constipation, foggy brain, amnesia, stunned look, or downright dopiness. You are not alone. The medical field calls it circadian dysrhythmia, but we all know it as jet lag. This feeling occurs when your built-in clock is out of sync with the outside environment, upsetting your bodily functions. One pilot's wife coined it the "dopey days." This foggy state has been blamed for athletes losing competitions, politicians blundering international relations, and businesspeople failing to secure major deals. Most people, including flight crews, succumb to it, but most aircrew grow accustomed to it. I am frequently asked how I deal with this travel haziness. I quickly retort: exercise!

Sadly, there is no magical fix, but there are certainly ways to miti-gate it. One suggestion from many aviation doctors is to get plenty of rest before the flight. Many passengers think they can catch up on their sleep during the flight, but it doesn't always work out that way. You may have a chatty seat partner, an unsettled child directly behind you, or a bumpy flight, or maybe you are not able to sleep upright. No wonder neck pillows are such a hit. Flight doctors recommend taking a nap for one and a half to two hours in the afternoon before a night flight, which is a low period in your circadian rhythm. On the day of travel, show up early, wear comfortable clothing, including good walking shoes, and have everything you'll need such as tickets and passports readily at hand. Airports are getting larger and more congested so they present a challenge to passengers who are pushing the time envelope. A stressed traveler is not a rested traveler.

During flight, get as comfortable as possible. Take off or loosen your tie, loosen your shoes, or better yet bring slipper socks. And even though airlines tell you to "store larger items under the seat in front of you," I would be rethinking that one if it's going to be a long flight. You want that option of stretching.

Everyone has been told to avoid alcohol because it is a diuretic, but that can be difficult. I like to imbibe a drink or two when I travel as a passenger. After all, often you are in an airplane because you are on vacation and in a fun place in your life and why not celebrate! But staying hydrated is a major key to warding off jet lag. Consuming water is very important — so much so, flight crew are given an extra liter of water for every eight hours of duty.

If you see a pilot walking down the aisle on a longer flight for no apparent reason, don't be alarmed: they are just limbering up, and you should be doing the same thing. I usually walk up and down business class without disconcerting the passengers too much. If I experience heads looking up and eyebrows rising, I won't venture further rear-ward into the economy cabin. I fully understand seeing a pilot in the cabin during flight makes some nervous, and yes, I've been asked, "Hey, aren't you supposed to be flying the airplane?"

Some passengers want to stay awake during the flight to maintain their normal sleep schedule (or, if they are like me, so they won't miss the meal service). But it can annoy others when you have your window blind wide open or all your reading lights at full blast. I've heard of confrontations because someone wanted to keep their window blind open to stay on their time zone.

Generally, it is said, flying "west is best," as your circadian rhythm is extended. In the other direction, your circadian rhythm, or internal clock, tends to be compressed, making "east a beast." What do compression and extension really mean? Your day is lengthened flying west and shortened heading east, and it's easier to stay up a few hours later. For me, this holds true. Those eastbound flights over the Atlantic are tougher than the ones heading west to the Orient.

Studies have clearly demonstrated that the most successful technique for combatting sleepiness and jet-lag-induced doziness is physical activity. I always have my gym gear packed no matter where I travel. You can also keep alert by reading, playing a game, or conversing. Going for a walk also works very well. Again, this is all personal preference. What works for me may not be so helpful for you. You must find your jet lag mojo.

Once you arrive, your ideal sleep schedule depends on the length of stay in the new time zone. If you are staying for only a day or so, as in the case with most crew layovers, it's better to keep close to your home sleep schedule. But if you are staying longer, heed the motto "When in Rome, do as the Romans do." That includes changing your watch to local time. I always change my watch to local time, no matter if it's one time zone over or on the other side of the world. That way I don't err in giving the wrong time in my PAs, and it prevents me from wrongly setting my alarm or missing my crew pick-up. It annoys me a tad when someone reminds me it is three a.m.

back home as we walk the streets in bright afternoon sunshine on the other side of the planet.

Crew have been advised by company doctors, if you can't fall asleep in 20 to 30 minutes, you should get out of bed so as not to associate bed with sleeping problems. Get up and read or watch TV. Incidentally, you don't need one long sleep. Two separate sleeps can be just as refreshing.

And don't use alcohol as a sleeping aid, because it upsets your natural sleep patterns. I should heed my own advice with this one, but I enjoy having a "sociable" with the crew. For me, it's part of the pilot package and helps make my job fun.

Having returned to long-haul flying, I get to hear the crew's personal take on how they mitigate jet lag. As well, I fly with very senior first officers who have chosen lifestyle — better sleeping patterns, better eating arrangements, and regular bowel movements — over promotion. Many pick the exact same routes and stay on their built-in clock. And believe me, lots of conversation is spent on their travel techniques. Sometimes it wears me out listening to all the various tactics. I'm also wondering, why did some of these people get into this business if they are so irritated with travel? It's like a doctor or nurse unable to handle the sight of blood. Some stay up all day, walk around in a daze, and eat when they want. Translation: "I'm doing my own thing on this layover, and I will see you at crew pickup." The word "boring" comes to mind.

> ✈ I know many airline employees who take sleeping pills and they swear by them. In fact, they are so adamant about it that on the crew bus I get to hear some of their wild, sometimes erotic dreams. I tried this route, but sadly my dreams were nothing to brag about.

The best way to ensure a sound sleep is to optimize your sleeping environment. Set a comfortable temperature, preferably on the cool side. Darken the room. Some pros bring along clips to ensure the curtains are drawn taut. Someone recently told me the hotel closet hangers with clips also work. Wear a sleeping mask if necessary and shut out any noise by wearing earplugs. Some crew members refuse to take rooms near elevators or ice machines. It is true that hotels find aircrew sometimes very demanding, but aircrew know what to expect and demand it.

I've had restless sleeps because my room shared the same floor with a rowdy sports team going from room to room or socializing in the hall, maids slamming the doors or vacuuming, people talking loudly in the next room or TVs turned up to near max, possibly because of jet-lag-induced hearing loss. I've been woken or couldn't sleep because of snoring next door, unsettled kids or pets, and couples loudly copulating through the night. It's not always easy for those on the road. Many think we airline types can morph ourselves into a sleeping cocoon. Some can, whereas many can't. I tend to be in the latter camp.

> ✈ I scratch my head when passengers start pouring out their travel woes. "I've been up since five a.m. and traveled all day." "This is my third flight today." "I've been up all night." Then they are taking selfies with their hair ruffled and posting it in the "poor me" category. We airline types just smile and want to say, "You are preaching to the choir."

Fatigue is a huge physiological parameter in the aviation world. In fact, fatigue is the F-word in aviation. We have rest rules that are dictated by aviation authorities as well as contract constraints that try to reduce and control fatigue. But sometimes it is hard to administer and just as difficult to adhere to. Everyone thinks aircrew get the proper amount of prone rest, but they are like everyone else: they have mortgages, owe bills, are in stressful relationships or away from family, have sick kids, are missing an important life event, and have every other issue that the traveling public have. The traveling passenger assumes everyone is fit, perky, ready to go, and at 100 percent capacity. All I can say is, we try our best.

WHAT DO CREWS GET UP TO ON LAYOVERS? WHAT IS THE HANKY-PANKY FACTOR?

Everyone assumes that promiscuity is part of the job, thanks to Hollywood's portrayal of pilots and flight attendants. Envision the slim young pilot depicted by Leonardo DiCaprio in the movie *Catch Me If You Can*, wearing cool sunglasses, flight attendants draped on both arms. And look at the sexy portrayal of flight attendants wearing tight outfits to entice businessmen. Sure, there are upstart charter companies trying to woo the public with nifty outfits and slim, fit people. Remember Hooters Air based out of Myrtle Beach, South Carolina, which lasted three years, from 2003 to 2006? Some airlines have a separate low-cost division that sprouts from their mainline, composed of younger flight attendants hired for their youth and vibrancy. And yes, I have heard stories. As well, the connector airlines generally have younger people, so things may be a little more robust on those layovers.

I have seen how jet lag, alcohol, and being on the other side of the world affect people's judgment. But most layovers are comprised of going out to find dinner with your flying partner, with little or no interaction with the flight attendants. When we do convene, someone inevitably makes a comment on the rendezvous rarity. Going out as a team seems to be happening less and less. Maybe it's me?

> ✈ Most of the time, I meet the "back end crew" when boarding the crew bus heading to the layover hotel. Numerous times, the question "You were on my flight?" arises. Some airlines make it mandatory for the pilots and flight attendants to meet and shake hands prior to the flight.

But you get co-workers partaking with other co-workers anywhere. Look at doctors and nurses, lawyers and secretaries, the entertainment business, or any office, big or small. "Dipping the pen in the company ink" is a fact of life. I am told most affairs happen when the aircrew is away and having trouble at home. And the stories are not always pretty. One pilot came home after a four-day pairing only to discover his wife had moved with the children to the other side of the country and put the entire move on his credit card. The house was barren; the only reminder of the kids was their smudge marks on the patio door.

AIRLINE ADAGES FOR RELATIONSHIP COHESION.

- Always call home if your flight is canceled or early. There have been many pilots over the years who didn't and came home to discover that their spouse or partner had intimate friends.
- The code word for someone visiting a "friend" on a layover is "cousin" as in, "I'm going to visit my cousin on this layover."
- "Stay married" is the best financial advice every pilot hears.
- If you are married to a pilot and they constantly fly to the same destination, then maybe an eyebrow should be raised.

Some countries look at pilots as demigods. You won't find this reverence in North America, but in other places of the world you will.

Even though many are middle-aged plus, bald with glasses, and can't button their tunic because of a bulging waistline, they look very desirable to many. For many locals, a pilot is a ticket to a better life. I know of specific countries (I am not telling you everything) where pilots had mistresses, wives, kids, and second families, so it was the only place they flew. If you find this difficult to grasp, then grab a copy of the bestselling book *The Pilot's Wife* by Anita Shreve. There is even a TV movie adaptation depicting how some pilots live double lives. But again, it's not only in the airline business.

Aside from the public's perception about what aircrew get up to on layovers, there are some incredible opportunities to see great spots and experience things most will never have an opportunity to. Here are just a few of mine: swimming in the Dead Sea, playing tennis in São Paulo, climbing Hong Kong's Victoria Peak, boating on the Rhine River, running through Vancouver's Stanley Park, sampling wines in Santiago, gorging on gyoza in Japan, and biking across the Golden Gate Bridge in San Francisco. How many jobs offer this diversity? But yes, there can be a jealousy factor with your significant other, neighbors, family members, and friends. If you want to be a pilot or flight attendant because of the so-called perk of promiscuity, you may be disappointed in a hurry. Most are in the airline business for different reasons.

> ✈ If only I made as much money people thought I made, had as much time off as my neighbor thinks I have, and had as much fun on layovers as my wife thinks I have, I would be a happy man.

"CALL HOME!" TWO WORDS NO ONE WANTS TO HEAR.

No one wants to hear "call home," especially if you just landed on the other side of the world. This very thing happened to my first officer on a recent flight to Paris. As we finished the shutdown checks, I opened the flight deck door, and a company representative was waiting to tell the first officer to call the duty pilot. It was the wee hours of the morning in North America, but the delegated weekend duty pilot had bad news. A member of the first officer's family had taken gravely ill, and things were not looking good. The first officer left on the return flight, with a car waiting for her when she arrived. Operations had to scurry and fly an alternate first officer from Frankfurt to Paris for the next day.

THE BEGINNING OF TIME. WHAT THE HECK IS ZULU TIME?

Pilots, air traffic controllers, and aviation weather specialists converse in an international time-zone language called Zulu. Each of the 24 time zones is assigned a phonetic letter. Z, or Zulu, denotes where time starts: 0° longitude, on a line that bisects Greenwich, England. We also use the 24-hour clock, so three p.m. in New York is 15:00 local but 2000Z (Eastern Standard Time) and 1900Z (Eastern Daylight Time). Zulu time, formerly GMT (Greenwich Mean Time), is now called UTC (Coordinated Universal Time). Though the ABCs of "Zulu" may be tough to grasp at first, the method keeps aviation and its weather in sync. To confuse the issue, Saskatchewan, Arizona, and Hawaii do not observe Daylight Saving Time. Some places, like Newfoundland and India, base their clocks on the half hour. Try giving a PA to passengers on the arrival time in New Delhi near the end of a 14-hour flight.

In 1878, Sir Sandford Fleming, a Scotsman who emigrated to Canada, proposed the system of worldwide time zones we use today after noting the inconsistencies of time implemented by the railroad system. Most towns had their own local time based on when the sun peaked at high noon. Fleming recommended that the world be divided into 24 time zones, since the Earth turns one full rotation every 24 hours. Though heralded as a brilliant solution to a chaotic problem, Fleming's time zone plan turned out to be difficult to implement, because each country wanted to be in possession of the "Prime Meridian of the World" — the place that the rest of world references when establishing time. After much debate, the Prime Meridian Conference selected the longitude of Greenwich as 0° and established the 24 time zones starting from Greenwich. Universal Time, based on the mean solar time in Greenwich, emerged and became known as GMT, or Greenwich Mean Time.

However, in 1972, GMT gave way to UTC, which uses the much more precise cesium atomic clock to keep time. The atomic clocks consider the tiny hiccups in the Earth's rotation of about one second every year by incorporating leap seconds. But most time zones continue to compute their local time referencing the Prime Meridian. Meanwhile, GMT no longer exists as a time standard, although the term GMT is often incorrectly used to denote universal time.

Just as determining the location for the epicenter of the Prime Meridian proved difficult, finding a label for the UTC met with resistance as well — hence the peculiarity of "UTC" standing for "Coordinated Universal Time." English and French speakers each wanted a term that reflected their respective languages: "CUT" for Coordinated Universal Time and "TUC" for Temps Universel Coordonné. This resulted in the final compromise of UTC.

ZERO HOUR.

Keep in mind that the baseline of the international time zone system is still the Prime Meridian, the line of longitude defined as 0°. Currently it's in Greenwich, but it has been in (or people have advocated for it to be in) Paris, Philadelphia, and near the Great Pyramid of Giza, among other locales. The modern Prime Meridian goes south from the North Pole through the United Kingdom, France, Spain, Algeria, Mali,

Burkina Faso, Togo, Ghana, Queen Maud Land (Antarctica), and on to the South Pole. A laser projecting from the Royal Observatory structure in Greenwich marks the line.

Greenwich itself is a popular tourist destination. It includes the Royal Observatory, perched on a hill overlooking the River Thames, and the Shepherd Gate Clock, which was the first to provide Greenwich Mean Time to the public. The clock is unique in that the hour hand goes around the dial once every 24 hours, so that at noon it's pointing to the bottom.

Atop the observatory is the conspicuous, red-painted Greenwich "time ball." The ball has dropped every day since 1833 at precisely one p.m., serving as a visual cue to the navigators on the River Thames to synchronize their clocks. In comparison, overhead satellites send continual signals to update flight deck clocks. Why not drop the ball at noon? Well, the astronomers chose one o'clock because at noon they were too busy with their astronomical duty of measuring the sun as it passed the local meridian. The ball-dropping is somewhat uneventful. It rises half way to the top at 12:55 and reaches the top at 12:58, dropping exactly at one p.m., but without noise — so be careful! With a blink of an eye, you may miss it. If you're planning to make a tourism stop, make sure you plan to get there before one o'clock local so you can see it drop.

DESTINATION UNKNOWN.

If you're traveling by plane, there's a good chance you're on your way to a place you've never been before. Though the prospect is exciting, you may have a few worrisome questions: Will the customs procedure go smoothly? Will the language barrier cause problems? And how will I find my way around the airport?

Pilots also land at airports unfamiliar to them but, luckily, protocols keep our anxiety to a minimum. For one thing, with few exceptions,

English is the universal language for both pilots and air traffic controllers, a bonus for native English speakers. Standard terminology is used to ensure information is communicated clearly and safely. Besides learning the industry terminology, any pilot must go through a qualification process before they can fly overseas. The process involves reading briefing notes, listening to a lecture on survival equipment, and undergoing a computer-based training program, along with a simulator exercise made up of oceanic and diversion (emergency) procedures. Once completed, the training is followed by a flight test in which a check pilot (supervisor) ensures that all the protocols have been understood and followed. Though the qualification process for domestic routes is slightly less rigorous, it still requires the pilot undergo a minimum number of supervised flying hours and a final flight with a check pilot (check airman).

In addition to this training, pilots flying to new destinations can consult their iPads, where they will find information on local weather and runways, as well as any special procedures required for an airport. It's a travel guide for pilots. Throughout my career, I've flown to more than 130 airports. Hopefully, you find landing at a new destination as exciting as I do.

> ✈ As I write this, I am on yet another deadhead flight. I was supposed to command a flight to San Francisco, but at the last minute, crew sked (scheduling) called stating the flight was "subbed" (substituted) and I was to deadhead. I met the new-hire first officer working the flight. It will be his first time to San Francisco. This unique airport requires either the captain or first officer to have flown there within a year, and if not, a detailed self-briefing is required. *What could go wrong?* I thought.

CHAPTER 8

WEATHER STUFF

CAN CLOUDS FORETELL FLIGHT CONDITIONS?

Clouds do foretell flight conditions. Billowy, puffy clouds signal a bumpy ride ahead because the air is unstable. Clouds that form a few thousand feet above the ground will typically give way to smooth conditions after you climb above their tops. Usually, layered clouds also lead to a steady ride. An experienced eye can recognize clouds or even fog associated with certain weather phenomena. The shape of the streaks that form from another aircraft's exhaust at higher altitudes can tip off pilots to flight conditions. I'm always attuned to the telltale signs of clouds, but that could just be the meteorologist in me.

Knowing how clouds form reveals the process transpiring in the atmosphere. Many think the nomenclature of clouds is extensive and difficult. Not so. By learning your clouds, you are indirectly learning flight conditions.

CLOUDS IN MY SKY.

In 1802, Englishman Luke Howard, a pharmacist by trade, gave clouds the common names we still use today. He noted three basic shapes of clouds: heaps of separated cloud masses with flat bottoms and cauliflower tops, which he named cumulus (the Latin word for "little heap"); layers of clouds like blankets, much wider than they are thick, which he named stratus (Latin for "layer"); and wispy curls, like a child's hair, which he called cirrus (Latin for "curl"). To clouds that generate precipitation, he gave the name nimbo/nimbus (Latin for "rain").

There are low, middle, and high clouds, and clouds of vertical development. Constantly examining the shape and intensity of clouds, and surmising what process formed them, helps me find smooth rides.

Clouds are composed of liquid water droplets, supercooled water droplets, ice crystals, or all of the above. Those high, thin, wispy clouds called cirrus, a.k.a. mare's tails, indicate fair weather, but if they start to thicken, inclement weather is approaching. They are composed entirely of ice crystals. Flying in cloud below zero Celsius may cause ice to form on an aircraft. Water can exist in liquid form in clouds well below freezing — in fact, all the way down to a frigid –40°C (–40° F — yes, Celsius and Fahrenheit are the same at this value). At that point, meteorologists coin it "homogeneous nucleation."

The idiom "being on cloud nine" is misleading. The billowy, lofty, puffy, cotton-batten-like cloud labelled number nine on the cloud chart — cumulonimbus — looks pleasant. But it's a wolf in sheep's clothing, and pilots know to avoid it at all costs. Think thunderstorm. Due to daytime heating, small cumulus clouds form and reform in about 10 minutes. Try watching it transpire with time-lapse photography, or better yet, lie on your back on a warm summer's day and watch it all happen.

UNDER PRESSURE.

We live at the bottom of an invisible ocean of air called the atmosphere. Though much of the time you can't see it, this ocean has many properties that must be considered when operating an aircraft.

This ocean is compressible and expandable, and yet it has immense weight and is marked by changes in temperature with height, air pressure, density, and oxygen level.

The higher you fly in the atmosphere, the colder it gets, because the Earth, warmed by the sun, indirectly heats the air. Also, when you ascend, air pressure decreases quickly. The outside air pressure while you are watching a movie well into the flight is about 20 percent of that at sea level.

✈ Many recall the Greek fable of Icarus and his father, Daedalus. In the fable, Daedalus warns Icarus not to fly too close to the sun, because his wings are made of feathers and wax. Icarus defies his father's wish, however, and as he nears the sun, Greek lore has him crashing back to Earth because the sun has melted his wax wings. We all know now that the logic here is erroneous: air gets colder with height, not warmer.

A third atmospheric property, density, also decreases with height; thus oxygen is extremely sparse because air molecules are farther apart. The pressure required to create a sea-level pressure at 35,000 feet (10,668 m) would require an airplane to be built like a heavily laden metal submarine. So why are we flying up at great heights anyway? When the air is less dense, aircraft encounter less resistance. Higher altitudes also allow aircraft to soar over most weather to give a smoother ride. Jet engines are more fuel efficient the higher they go; by producing less thrust, they burn less fuel, so it's imperative we fly higher, generally.

Pressurized cabins arrived in the early '40s with more streamlined cylindrical designs. Aircraft fuselages have retained this shape, which ensure a uniform level of pressurization. To pressurize an aircraft, air from the engines, called bleed air, is taken via ducts from the compression chambers of the jet engine. This hot air must be cooled first and then enters the cabin and cargo holds. To have a continuous flow, some air must be vented to the outside by a set of outflow valves, with the remainder of air recirculated.

WHY THE BUMPS?

Over 24 years ago, my first article for *enRoute,* titled "Why the Bumps?," explained the seven types of turbulence. This is the most commonly asked question and concern from passengers. Technology has certainly helped to contend with the unpredictable nature of bumpy rides. A pilot's iPad can now superimpose a specific routing on weather charts that depict areas of bumps. We can also interrogate other airplane reports from this site. These "heartbeats" show the whereabouts of the airplane, its altitude, and whether bumps occurred during the flight. Our flight plan also assigns a numerical value for possible bumps along every waypoint, while flight dispatch sends us in-flight reports via datalink. Everyone is striving to deliver the smoothest ride possible.

TURBULENT TIMES.

Physicist Theodore von Kármán is credited with saying, "There are two great unexplained mysteries in our understanding of the universe. One is the nature of a unified generalized theory to explain both gravity and electromagnetism. The other is an understanding of the nature of turbulence. After I die, I expect God to clarify the general field theory to me. I have no such hope for turbulence."

My flight plan assigns a number from zero to nine to each navigation point along the route. A zero or one means things should be smooth, but when numbers three, four, and five appear, the seat belt sign is likely to be illuminated. Many think there is a device that automatically turns on the seat belt sign when bumps occur, but it's strictly subjective and done with the concurrence of the captain.

✈ Finally, the internet has made its way into the flight deck. For a few years, for technical reasons, the internet was available only in the cabin. I would joke that I could approach a passenger in business class during the flight and ask him what the weather is doing up ahead. Now, I don't have to.

IS THERE ANY WAY TO DETECT SEVERE TURBULENCE AND DETOUR AROUND IT?

It's policy and common sense to avoid known areas of severe turbulence, especially thunderstorms. Most airliners also have wind shear systems to detect shearing winds near the ground. These do not detect high-level wind shear. No device detects turbulence due to jet streams, but weather maps depict and forecast all types of turbulence. Flight dispatchers will plan for flights to avoid these areas or fly at different altitudes. Sometimes this is all it takes to ensure a smooth ride.

✈ Many innovative companies, including aircraft manufacturers, have been experimenting with LIDAR, a form of turbulence detection. LIDAR (Light Detection and Ranging) uses light from a pulsed laser to measure ranges to a target (swirling air). As of now, this forward projection from the aircraft is distance-limited, giving a pilot a heads-up of perhaps 60 seconds. If anyone out there can build a system to detect that elusive turbulence earlier, you will be a billionaire.

Having said that, some areas of turbulence cover a huge swath. Some days we are up there trying to find smooth air, but the seat belt sign won't budge to the "off" position. Air traffic control, inundated with ride requests, will describe the scenario as "bumpy at all altitudes." To avoid some turbulent areas would require deviations of hundreds of miles, adding tens of minutes or even hours and possibly requiring more fuel than the tanks can hold. To run an airline on schedule takes a ton of juggling, and trying to keep things smooth is part of the challenge.

✈ Dear Mr. Morris: Whenever I fly, the second I sit down I reach for the *enRoute* magazine and your article of the month. I am a very, very nervous flyer, often to the point of tears. I often want to drive to Vancouver and even risk the Coquihalla [Highway] in the winter. My fear of flying has greatly impacted our travel plans, much to my husband's dismay. And, yes, it is usually associated with turbulence. Therefore, I was very happy to see this article. I live in Kelowna, B.C., and most of the time I am leaving here it is on a prop airplane. It isn't the prop I mind so much, although I would prefer a jet, it is the fact that almost all the time coming in and out of Kelowna, there is turbulence. Not just a bump or two, often I fear we are going to fall from the sky it is so rough. And often, the weather appears just fine. So, I don't get it. Having said this, the last time I flew one and a half weeks ago from Vancouver to Kelowna, the captain made an announcement prior to takeoff to say that it would be rough going into Kelowna, but nothing to worry about. Just hearing his calm announcement and his reassurance that this was all quite normal helped me tremendously. I can't tell you the difference it made in my short flight home. That is only the second time I have ever heard from the cockpit in many years. Usually there is nothing said at all. When I got off the flight, the captain was standing in the cockpit doorway and I told him exactly how much I appreciated what he had said. It took so little time and meant so much to me and I am sure others.

— Susan

IS TURBULENCE DANGEROUS?

No, and sometimes yes. Any airliner must be built to handle large amounts of turbulence, but no sane person wants to challenge this extreme realm. Yes, you will see the wings flex up and down as they contend with rough air. This is normal, although for most passengers it is disconcerting. The danger arises when passengers don't have their seat belts fastened, as rough turbulence can appear suddenly, catching many off guard. It can toss you about violently, and if items are not secure, they can act as projectiles. People try to get comfy by undoing their seat belt or allowing a child to sleep on the floor. They are the ones injured when Mother Nature starts throwing nasty punches. Over my entire career, I can count on one hand — okay, maybe two — where things got scary rough. I admit, I am a wimp when it comes to rough air, and for you as a passenger this is a good thing.

Many passengers get a false impression when the seat belt sign is illuminated and yet the flight attendants are going about their business. Some pilots turn on the seat belt sign at the onset of the first ripple, envisioning lawsuits if they don't illuminate the sign. Deciding when to put the sign on is purely subjective. For a long-haul flight the seat belt sign may cycle on and off 10 or more times. This on-again, off-again trend can foster a laissez-faire approach among passengers.

And don't think pilots are up in the flight deck saying nonchalantly, "Oh well." We don't like rough air either. It wears on everyone.

✈ Part of our mandatory briefing with the in-charge flight attendant includes reporting on expected ride conditions. One day, the in-charge flight attendant quickly intervened, saying, "I know, I know . . . it's going to be smooth between the bumps."

DESCRIBING TURBULENCE IS AKIN TO DESCRIBING THE TASTE OF WINE — IT'S SUBJECTIVE.

Not only is turbulence hard to detect, but it evades precise description, and I find it challenging to convey when teaching pilots. It's like describing the taste of wine — purely subjective. Four intensity levels exist: light, moderate, severe, and a category some weather books don't acknowledge, extreme. Intensity is predicated on aircraft size, pilot experience, and even the mood of the pilot. A pilot commanding a large twin-aisle Boeing 777 may describe the bumps as light, but a newly licensed 18-year-old in a "bug smasher" may describe it as the worst turbulence ever. I could bore you with how the governing authorities define it, but it will only induce bewilderment. We also depict its frequency: occasional, intermittent, and continuous, and we distinguish between "chop" and turbulence. Chop is rhythmic, whereas turbulence is more irregular. Some may find this diagram silly, but I think it describes things nicely.

TURBULENCE INTENSITY

LIGHT	Coffee Sloshing
MODERATE	Coffee Spilling
SEVERE	Flight Attendant Spilling
EXTREME	Aircraft Spilling

All right, here is my way of describing merlot-to-shiraz ride conditions. Light bumpiness is annoying. If it persists, the seat belt sign will probably be illuminated, but it depends how far the pilot will go with an imaginary lawyer in their decision process. And if the "lawyer in their ear" is not enough, a flight attendant in the back of the airplane will call the flight deck reminding the pilot the tail of the airplane is swinging more and politely ask to have the seat belt sign turned on.

When the bumps increase from jiggles to jolts, we're entering the realm of turbulence. That's when a pilot comes out of their happy stupor, especially when they hear other pilots on the radios describing the bumps as turbulence instead of chop. The seat belt sign is on, and hot drinks will not be served. Flight attendants are still up in the aisles but warn passengers they better think twice about heading to the washroom. Pilots are asking about ride reports and thinking about a flight level change or possibly slowing a bit to reduce the impact of bumpy air.

With moderate chop and/or turbulence, passenger conversations stop. They are now looking at the flight attendants for answers. Their faces start changing expression. Some start clenching their hands. Flight attendants are now returning their carts and securing the cabin. Things get quiet in the cabin. Pilots are talking curtly on the radios to find smooth air and slowing the airplane to rough-air maneuvering speed to ride out the bumps, like a boat slows when waves strengthen. Lavatory doors are rattling, dishes and cups are clattering in the galley, and filling out your customs form is impossible.

Now we enter severe. That's when you may hear a scream or two — and not from the flight attendants. (There is no such thing as severe chop.) Pilot voices have escalated several octaves on the radios. There will be paperwork to fill out. And sadly, this is when social media becomes abuzz. I better stop.

In my experience, about 85 to 95 percent of bumpy air is of light intensity. Bumps of moderate intensity happen less than 5 percent of the time, and severe will rear its ugly head once in a blue moon. But remember, most flights are predominantly smooth, with the odd jiggle.

✈ The Boeing B787 I fly has a gust suppression system. Strategically located sensors send signals to the flight controls (rudder, elevator, spoilers, ailerons, and flaperons) to neutralize the bumps. Several passengers, about 8 percent, succumb to motion sickness, a number that the system reportedly reduces to 1 percent. Airsickness bags begone!

DID I HEAR THERE ARE SEVEN TYPES OF TURBULENCE? I THOUGHT BUMPS WERE BUMPS?

Yes, agitated air is agitated air, but here are its causes:

1. Convective turbulence (up-and-down motion) is due to daytime heating from the sun, peaking mid-afternoon. Thunderstorms are the violent result of intense daytime heating. Morning, late evening, and night flights tend to be smoother as they avoid convective turbulence unless substantial cloud developed from the heating process.

2. Mechanical turbulence is due to strong gusty surface winds blowing over the terrain and built-up areas. If it's windy, there is a good chance of bumps at low levels. The good news is it smooths out fast just a few thousand feet above the ground. Convective and mechanical turbulence frequently merge to give a rock-and-roll ride. During approach, we will slow the aircraft. On departure, we steepen the climb to mitigate the jolts.

3. "Orographic" is a big word for turbulence forming from wind flowing over a mountain or even a hill. The Rockies, the Andes, and the Alps dish out undulating mountain waves. It may be difficult to get above the mountain wave, and if unable, it may require a steeper or quicker descent. Pilot reports are invaluable in the decision making.

4. LLWS (Low Level Wind Shear) is a sudden change of wind speed and/or wind direction. This can occur under a thunderstorm, from the funneling of winds down a valley, or from strong surface winds, to name a few causes. Most large airports in the U.S. have wind shear detection. Canada doesn't. As well, all airliners have wind shear detection on board. But wind shear can vary from little to no turbulence

to sometimes violent bouts. LLWS came to the forefront during a Delta airlines flight in 1985 on approach into Dallas Fort Worth, Texas.

5. CAT (Clear Air Turbulence) is a high-altitude rough ride usually associated with jet streams — but not always. It's a bit of a misnomer because substantial cloud may be present. CAT is mostly a cruising-level occurrence. This can go from smooth to very rough within seconds, catching many passengers off guard with their seat belts unfastened. Keep it on — to keep it safe! CAT causes the most injuries, with flight attendants receiving the brunt of it.

6. Frontal turbulence is due to surface fronts and their associated wind shifts.

7. Finally, there are torrent eddies from other airplane wing tips, called wake turbulence. Generally, the bigger the airplane the bigger the chance of spreading rough rides to other airplanes, like a large ship fanning out strong waves to other boats in its wake. Because navigation is so precise, we frequently are lined up laterally to within feet. There's no better example of this than flying over the oceanic routes. We use an offset procedure called SLOP (Strategic Lateral Offset Procedure) that allows us to displace one or two miles to the right. A "slopping" pilot is a safe pilot.

DOESN'T WEATHER RADAR DETECT TURBULENCE?

Onboard weather radar detects precipitation, which, if significant, implies turbulence because the air is unstable and moving about to cause a bumpy ride. The nose, or radome (radar dome), of every airliner is where the weather radar and antenna is fixed. The radome can be hinged open on the ground to inspect the weather radar. Weather radar, derived from World War II technology, is a great detector of rain; however, other types of precipitation reflect less energy back, so a pilot must challenge these "returns." Generally, light precipitation is shown on the radar screen as green, heavier rain is yellow, and heavy to extremely heavy is red. Some radars use magenta for the extreme category. Pilots want to avoid all returns. Red should be avoided at all costs — this is no place for any airplane, and pilots know it. We want to avoid, avoid, avoid — but sometimes pilots get into tight situations and things can get rough. Rest assured, the seat belt sign will be illuminated. One recent frequent flyer, nearing the million-mile category, asked if air traffic controllers purposely vector pilots into bad weather because of constraints. They will try their best not to, but sometimes a pilot must weigh the consequences. This is not a good predicament, and it's not where anyone wants to be.

✈ One of the best methods, still to this day, for avoiding showers, heavy rain, and turbulent cloud is with a pair of eyeballs. At night, you'll find me with the flight deck lighting turned down, looking intently outside and scanning the sky. Many weather books geared for pilots tell the pilot to turn the lights in the flight deck to full brightness so that the lightning outside is less blinding. I say hogwash! We are near a thunderstorm, so I have the flight deck lights dimmed and, usually, the external strobe lights turned off, finding the best path to get out of this meteorological predicament.

WHAT'S A JET STREAM (ATMOSPHERIC SKY SNAKES)?

We all probably have some vague notion of what a jet stream is: those streaky white trails that appear behind high-flying jetliners?

Well, no, that's a common misconception. Those are condensation trails (contrails) formed by the exhaust of aircraft jet engines. Jet streams are a much more spectacular phenomenon — long, thin bands of extremely fast-moving air that form at high altitudes and corkscrew through the atmosphere around our planet.

In 400 BC, Aristotle wrote a treatise on weather entitled *Meteorologica*, in which he noted that higher clouds may move faster than lower clouds. Over 2000 years later, when manned balloons were first launched during the 18th century, those aboard noticed that winds tended to increase with height. But it wasn't until World War II and the advent of high-altitude flight that jet streams were encountered and their presence confirmed.

On November 24, 1944, 111 U.S. Air Force B-29 bombers were sent from the Pacific island of Saipan to attack industrial sites near Tokyo in the first high-altitude bombing mission of World War II. As the airplanes approached the island of Honshu at 33,000 feet (10,058 m), they were suddenly hit by winds of 140 knots (161 mph or 260 km/h), which knocked them completely off course. Only 16 of the 111 pilots managed to hit their targets, while the rest were blown over the ocean and forced to return to base. Pilots on subsequent missions also reported encountering extremely powerful winds and unexpected turbulence when flying westward to Japan. What were these incredibly strong winds? One pilot likened them to a jet of air streaming out of a hose with enormous velocity; hence the name "jet stream."

We now know that jet streams are produced because of the significant temperature changes where air masses collide. In North America, for instance, where there may be four distinctive air masses, up to three separate jet streams would exist. They can be hundreds of miles long, tens of miles wide, and a few to several thousand feet thick. I liken their dimensions to those of a Christmas ribbon. They migrate southward in winter and decrease in altitude the further north they are found, and flow at different heights depending on the season.

Jet streams are strongest during the winter months because the frontal zones or temperature differences between air masses are more dramatic in winter. They mark the dividing line between seasonable and unseasonable temperatures. They also indicate in what direction and at what speed surface weather (highs, lows, fronts, etc.) is traveling.

To locate jet streams, weather balloons are sent up to penetrate the higher atmosphere, climbing to 100,000 feet (about 30 km). As well, airliners have equipment on board to gauge upper atmospheric wind conditions, and satellites capture some of the telltale cloud patterns associated with jet streams. This fast-moving current of air circulates around the Earth in a corkscrew motion at altitudes of 25,000 to 45,000 feet (7620 to 12,192 m). A jet stream's speed and momentum are affected by the varying temperatures in the southern and northern latitudes as well as by the Earth's rotation. Average high speeds are between 100 and 180 knots (115 and 207 mph or 185 and 333 km/h). I have pictures of flight deck instruments depicting speeds howling at 200 to 230 knots (230–265 mph or 370–426 km/h). The force of a jet stream often leads to quicker flight times by creating what pilots refer to as "push." A flight from New York to Los Angeles can take five hours, but with the effect of a jet stream, the return trip can be reduced by 40 minutes or more.

Sometimes a pilot will want to get into this fast-moving air; at other times, they will want to avoid it. We try to capitalize on strong tailwinds and, if able, duck out of the stronger-than-normal headwinds. If any turbulence is detected, pilots always ask for ride reports from air traffic control and then climb or descend to find smoother air. On occasion, this air can also be rough and become so quickly, which is why regulations have you keep your seat belt fastened at all times.

✈ Robert Buck, in his iconic pilot weather book *Weather Flying* (1998), described the elusiveness of high-altitude turbulence this way: "Meteorologists can locate the jet stream rather accurately, they cannot pinpoint exactly where the turbulence will be. They can tell you in general terms. I can assure you that sometimes it will be rough when they say it won't and smooth when they say it will be rough." To this very day, it remains a hard and fast observation.

AURORA BOREALIS — NORTHERN LIGHTS (DAWN OF THE NORTH).

Winter's long nights and frequent clear skies in northern latitudes render fantastic views of the dancing light show some 60 miles (97 km) above. Charged particles released from the sun (solar wind) collide with the Earth's magnetosphere. This causes earth's oxygen to yield a pale green or pinkish display. Higher up, about where the space shuttle cruised (200 miles), nitrogen produces blue or purplish-red. Many of our flights take in this awesome display, so keep your window shades open to catch Mother Nature's free light show.

Northern lights facts:

- The northern lights dazzle about the Earth's magnetic poles, so northern Canada (where the magnetic pole is located) is privy to some of the best shows, especially the Yukon, Nunavut, and the Northwest Territories. The southern lights are known as aurora australis.
- Santa Claus operates from the geographic North Pole, but the aurora borealis centers on the magnetic pole.
- On many northern flights, I've witnessed the many guises of the dancing northern lights — from docile patches of light, bursting streamers, arcs, and undulating curtains, to spiking rays sometimes with ominous glows.

- When Mother Nature sets up her stage full of lights, we frequently ask flight attendants to visit the flight deck to take in the show.

WIND BENEATH OUR WINGS.

Wind facts:

- Winds are reported and forecast in the direction of true north, but air traffic control at the airport state them in relation to magnetic north because runways are oriented to magnetic.
- I've been in winds clocked at 220 knots (253 mph or 407 km/h) at cruising level.
- The minimum speed to constitute a jet stream in Canada is 60 knots (69 mph or 111 km/h). In the U.S. and Britain, it's 80 knots (92 mph or 148 km/h).
- The device that measures wind speed and direction is an anemometer. Strictly speaking an anemometer measures wind speed, but colloquially the word is used for devices that determine both speed and direction.
- Wind is measured at a height of 10 meters (33 feet) at airports around the world.
- Winds are referenced to where they are coming from, NOT where they are going. A south wind means it is blowing from the south.
- Every airliner has a wind readout and groundspeed — how fast we are moving across the ground due to winds.
- Iconic singer Bob Dylan claimed, "You don't need a weatherman to know which way the wind blows." But in aviation you do!

- Information about upper-level winds (crucial for flight planning) is derived from weather balloons from over 900 stations around the world, launched twice a day.
- Data from aircraft, called AMDAR (Aircraft Meteorological Data Relay), is also used to drive the weather supercomputers forecasting winds at all flying levels around the world.

I NOTICED A CIRCULAR RAINBOW WITH THE SHADOW OF THE AIRPLANE INSIDE IT. HOW DOES THAT OCCUR?

A "glory" forms in moisture, and the larger and more uniform the droplets, the better. These droplets act like little prisms, scattering the primary colors and bending them back to you, the passenger. You need the sun behind you for the aircraft to create a shadow in the center as a bonus. When an aircraft's shadow is seen to dance inside, it's called "the glory of the pilot." Glory may be a precursor to possible airframe icing. Pilots don't like flying very close to cloud tops for two reasons: described as "bouncing on the tops," it implies a bumpy ride and potential airframe icing.

THE DEICE MAN COMETH — TAKING IT OFF AND KEEPING IT OFF.

As you sit comfortably in your seat, take a moment to notice ground crew working outside in freezing temperatures. They appear to be washing the airplane, but they're ridding the wings and sometimes the fuselage of ice and snow. This procedure is deicing and is necessary for departure, as airline safety standards prohibit takeoff when ice, frost, or snow is adhering to the airplane. But why?

Many think it's because those contaminants add weight, but the main issue is that they disrupt smooth airflow over the wings, which decreases lift. The captain is ultimately responsible, but the lead ramp attendant must also be in concurrence. Even flight attendants and passengers can voice concerns.

No place better illustrates deicing than the CDF (Central Deice Facility) at Toronto's Lester B. Pearson airport — the world's largest such facility. This 65-acre "drive-through airplane wash" consists of six huge bays that can be subdivided into three to accommodate smaller airplanes. Official deicing season is October 1 to April 30. Because this is a "live" or "engines running" operation, precise terminology and electronic sign-boards are used to keep things safe. Pilots contact the "Iceman" in the deicing control tower, appropriately named the "Icehouse." Once in position, two or more expensive ($1.4 million each) made-in-Denmark vehicles called the Vestergaard Elephant Betas (the facility has about 40 of them) spring into action. The deicing procedure involves spraying an orange deice Type I fluid composed of hot water and glycol to rid the airplane of ice and snow. If precipitation is falling or imminent, it is followed up with a cold application of a bright neon-green anti-icing Type IV fluid to stop precipitation from sticking. The "throughput" time for an Airbus 320 is an amazing 12 minutes. For a light snow event, it takes about 300 liters (80 U.S. gallons) of Type I at a cost of one dollar per liter, and 250 liters (66 U.S. gallons) of Type IV at two dollars per liter, plus a $350 visit fee — translation, about $1150. I've been deiced in Buffalo, New York, where the vendor soaked us, in more ways than one, with a $25,000 bill. This is all part of doing business in winter.

IS WEATHER GETTING WORSE? IS TURBULENCE MORE INTENSE? ARE THUNDERSTORMS BIGGER?

If you fall for the "CNN Effect," then you will be convinced turbu-lence is more frequent and more violent. I can't give a definitive yes or no, just a maybe. Mother Nature must balance the books, and that includes the heat budget. For jet streams to be stronger, temperature differences would have to be increasing — i.e., the North Pole would be getting colder and areas further south getting warmer. For thunderstorms to get bigger, the layer in which we live, the troposphere, has to be getting higher. I'm not so sure. But expect more articles to infiltrate the media. We're going to learn more terms and hear more opinions and surveys, as we did for the "polar vortex" and the "cyclone bomb."

✈ Patrick Smith, an airline pilot and a prolific writer, calls it the PEF (Passenger Embellishment Factor) in his best-selling book *Cockpit Confidential*. He states passengers have a way of over-describing turbulence, either from lack of experience or for effect. The media ups it a notch in the embellishment factor.

CONTRAILS OR CHEMTRAILS?

Man-made clouds may form behind an aircraft, produced by the moisture of combustion exhaust saturating the air and causing condensation. Two byproducts of hydrocarbon combustion are carbon dioxide and water vapor. For each pound of jet fuel burned, about 1.4 pounds (0.6 kg) of water vapor is produced. Many believe that the contrails we see in the sky are pollution, but they are mostly frozen water. The vapor condenses into tiny water droplets, which freeze if the temperature is low enough. These millions of tiny water droplets and/or ice crystals form contrails. The exhaust particles act as a trigger, causing the trapped vapor to rapidly condense. Exhaust contrails usually occur above 25,000 feet (7620 m), and only if the temperature there is below –40°C (–40°F).

Conspiracy theorists, akin to the flat earth society, are adamant that these white ice crystal streaks are chemtrails (chemical trails) imposing harm. It is said, "Don't believe everything you read on the internet," and this is a prime example.

OUR ATMOSPHERE.

Four main layers of the atmosphere exist. We live in the troposphere, where most weather occurs, but your flight may also take you into the second layer, the stratosphere. The boundary between the two layers is deemed the tropopause. Pilots abbreviate it as the "trop," which rhymes with "rope," not "top." This interface is known to be where turbulence lurks. It acts like a lid to most weather, more specifically thunderstorms, and is where jet streams coil around the globe. It's coincidentally the altitude where jet engines are most efficient. Because of it, pilots always want to know the whereabouts of the tropopause. It rises in the summer, is lower to the north, and rises in southern latitudes. It also changes day by day according to the weather systems. Statistically it hovers around 36,000 feet (10,973 m) but elevates to 55,000 (16,764 m) toward the equator, thus higher thunderstorms. A pilot's flight plan includes the location of the tropopause and is specified to within feet at every waypoint along the way.

HUMIDITY AND TEMPERATURE.

Most find it hard to fathom that moist air is less dense than dry air. You're probably asking, why is that? Since I can see moisture like fog, mist, and haze, shouldn't it be denser? Nope. Water weighs less than nitrogen, the main constituent of air. But temperature affects air density more than moisture. That's because there are fewer air molecules in a given volume of warmer air than in the same volume of cooler air. Thus, the air on a hot, humid summer's day is less dense than on a cold, dry winter's day. This thinner air plays into aircraft and jet engine performance because when thinner air flows over a wing, it means less lift, and thinner air in a jet engine means less thrust.

The humidity of cabin air in most airliners is derived from conditioned air ducted from the engines and has humidity levels equal to desert air (5 percent). The new B787 Dreamliner and the Airbus A350, however, provide a more humid cabin allowing a more restful flight with 15 to 20 percent humidity. On the B787, cabin air pressurization is provided by electrically driven compressors, eliminating the need to cool heated air from the engine. The cabin's humidity is programmable and based on the number of passengers carried.

FOGGY THOUGHTS.

An aviator will learn fog has many guises. In fact, there are six mechanisms whereby fog will form. One of the foggiest airports on the planet is on Canada's east coast: St. John's, Newfoundland, where visibility will drop to a half a mile or less for nearly one-third of the year. Fog will form when air moves up a hillside, as warm air moves over a cool surface; when cool air advects (moves horizontally) over warm water; in frigid temperatures, when it rains; and overnight under clear skies and light winds. Fog's less restrictive counterpart is mist. Weather observations abbreviate mist as BR (some refer it to "British Rain"), which stems from the French word "brume." Here is my poetic attempt to explain the six types in pilot verse.

Brume
It prances in diverse guises
It marks its misty presence as it ascends a hill little by little
As a warm wind moves over chilly waters it will form an immense
white blanket
It can stay for days and wilt a spirit

Or appear at dusk and ebb at dawn
It may accompany a gale, obscuring a pilot's line of sight to mere feet
It can ally with raindrops, inducing low visibility
Or play havoc in bitterly cold Arctic air
It can be a sign of seasonal change as it lunges from warm water
But no matter its genesis . . . it will challenge any aviator . . .
— Captain D

CAN IT BE TOO HOT TO FLY?

You'll be hearing more and more about this as we progress into global warming / climate change. Recently, many flights were canceled in southwestern USA because of the heat. Much of the southwest is higher in altitude, also contributing to thinner air. One saving grace is the air is drier. You now know dry air is denser than moist air.

Thus, it can be too hot to fly, as air is less dense and when it gets extremely hot, aircraft takeoff calculations will forbid safe departures. Every airliner in the world must have a balanced field in case of a rejected takeoff on the runway so it can come to a safe, complete stop. When higher speeds are required to produce enough lift to get airborne, this balancing act for takeoff speeds becomes a major player. One way is to reduce the weight, meaning fewer passengers or leaving cargo behind.

✈ Once, years ago in India, the temperature hovered at 31°C (88°F) near midnight — too hot for a safe and legal takeoff. When it cooled to 30°C (86°F), we were good to go. It still made for an interesting takeoff, with the aircraft laden with the fuel required for a 15-hour flight, and a full load of passengers.

You'll find many airlines in the Middle East operating most of their flights during the wee hours of the night, as temperatures are somewhat cooler. Luckily for them, most airports sit at elevations near sea level. On that note, airports at higher elevations can be problematic. Denver (the Mile High City), Mexico City, and El Dorado, Bogota (elevation 8360 feet, or 2548 m), offer challenges. Luckily for Denver, they have some of the longest runways in North America. It's also why Calgary, Alberta, has the two longest runways in Canada: it sits at an altitude of over 3600 feet (1097 m) above sea level, whereas Vancouver on the other side of the "Rocks" sits at 13 feet (4 m) above sea level.

One must also factor in ground operations. The heat can be dangerous to ground personnel, especially in the belly of the airplane. Animals could perish in the heat, so restrictions are implemented on pet transport. As well, airliners themselves have temperature limitations, as they must keep certain components cool, such as the aircraft avionics. If the on-board air conditioning units or supplied ground air can't keep up, then boarding will be denied.

✈ An exception: years ago, while we sat at the stand in London, England (they call gates "stands"), our air conditioning unit was not working. The captain refused to allow boarding until we had acceptable air conditioning, but none was available. However, London of all places does not have an issue with oppressive heat, and it was still safe to fly. Luckily, I convinced the captain to allow boarding, or else the flight would have been canceled.

CAN IT BE TOO COLD TO FLY?

Cold temperatures mean denser air, which is welcomed by any aviator.

Frigid air at −40 °C (−40 °F) is about one-third denser than hot air at 40 °C (104 °F). A flight departing Chicago O'Hare, gripped by bone-chilling air from the enigmatic polar vortex, will encounter about 33 percent denser air than a mid-summer departure in Dubai. Denser air produces more lift over the wings and flight controls as well as more thrust from the engines and propellers. But to start a jet engine requires oil temperatures above −40°C (−40°F), so the engines must be preheated. The airplane itself is accustomed to frigid temperatures at cruising altitude. It is the ground personnel who are challenged during extreme cold. They must retreat to the inside frequently, for safety reasons. Machinery is reluctant to start, heaters are less effective, and getting potable water to the airplane can be an issue, as well as cabin doors freezing shut. If an aircraft remains at the gate overnight, ground power must remain on to prevent water lines from freezing, and if it's parked away from the gate with no power, the water is drained. Even the coffee must go. The shipping of livestock is also challenging. Hairless cats and dogs are forbidden to travel during the winter.

During winter's extreme cold snap, I've seen the wheels to the jetway freeze, requiring the ramp attendants to take 20 minutes to thaw out the frozen wheels by blowing hot air on them. Aircraft parking brakes could freeze, so pilots may elect to release the brakes when the aircraft is safely chocked.

Winter operations are a challenge, but when the mercury plummets, any airline's goal remains to get you to your destination safely.

FRIGID FACTS.

When the temperature is 0 °C or below, a pilot must adjust their decision height to land by consulting a cold temperature correction table. For example, if the minimums for a landing (lowest a pilot can descend before they must see the runway) is 750 feet (229 m) ASL, but the temperature is −20 °C (−4 °F), then 20 feet is added so the new height is 770 feet (235 m) ASL.

For those pondering learning to fly, don't let the prospect of winter discourage you. Flight during a cold, crisp, clear winter's day can be a pleasant adventure. You'll understand what pilots mean when they describe the climb performance as resembling a "homesick angel."

Operations states flights will not be planned to operate for periods longer than 90 minutes in areas where temperatures are −65 °C (−85 °F) or colder. An aircraft can change altitude, avoid the area, or speed up.

Mercury thermometers freeze at −39 °C (−38 °F), so an alcohol-filled thermometer is used thereafter.

The magic temperature at which snow starts to squeak is −10 °C (14 °F). You may encounter this while walking to an airplane parked out on the ramp.

Water can exist as a liquid at temperatures as low as −40 °C (−40 °F). Thus an airplane flying through cloud can pick up airframe icing, as these supercooled water droplets lose their heat when they impinge on the airplane.

High cloud with bases starting above 20,000 feet (6096 m) is composed entirely of ice crystals because of the frigid temperatures aloft.

One place aircraft manufacturers take their planes for cold-temperature testing is Iqaluit, Nunavut, in northern Canada.

The freezing point of the Jet A1 fuel used by airlines is −47 °C (−53 °F). The rate of cooling for fuel is about 3 °C (4 °F) per hour. According to Boeing, an increase of .01 Mach — e.g., increasing from Mach .84 to Mach .85 — increases the airplane's skin temperature by .5 °C (.9 °F) to .7 °C (1.6 °F).

Square tires? Sometimes when an aircraft sits in prolonged cold temperatures, the aircraft tires take a while to warm up, so taxiing may be a bit bumpier. A few pilots have been known to mistake these square tires for flat tires.

One recent fast freeze at Toronto Pearson caused the covers to the fuel hydrants to freeze shut. Delays occurred until applied heat thawed things.

Two types of fog develop during frigid temperatures: steam fog emanating from open water, and ice fog from car and jet exhaust.

How does a pilot know frost is reported at a weather office? When FROIN (Frost On Indicator) is observed.

CAN IT BE TOO WINDY TO FLY?

Wind is the movement of air, and we need lots of moving air over the wing to fly. Remember, we are moving at about 110 to 170 miles per hour (177–274 km/h) into the air to get airborne. Pilots want to land and take off into the wind. Crosswinds can make it difficult or impossible to control the aircraft during takeoff or landing. But what about a 100-mile-per-hour wind blowing right down the runway? It is doable, but getting to the runway would be the challenge. Plus, you must think about ground operations and flying debris. I've seen baggage containers, technically called ULDs (Unit Loading Devices), blow around the tarmac like tumbleweeds. As well, the associated turbulence whipped up by the surface winds may keep the airplane grounded, or require an approaching aircraft to find another airport with more favorable winds.

IS LIGHTNING DETRIMENTAL?

Many envision those Hollywood B movies where an airplane is in a turbulent ride, flying in and out of dark ominous clouds, when suddenly a lightning bolt knocks part of the wing off. Lightning will enter the aircraft and exit with no damage — usually. However, it may leave little pinholes or burns during its transit. Statistics show an airliner gets hit every 5000 hours, or about once a year. The FAA estimates every airliner in the U.S. will be struck once a year. If we do get christened with an electric jolt, the aircraft must be looked over with a fine-tooth comb by maintenance. Aluminum is an excellent conductor, but some airplanes made of composite may experience lightning strikes a little more. Did I tell you I fly the composite B787?

✈ On my first flight flying solo as captain on the B787, the first officer looked over at me and mentioned the slightly higher probability of lightning strikes with the B787. Guess what transpired that same day on the return flight from Los Angeles to Toronto? I had to write up a lightning strike event in the logbook when we landed. Having said that, I've been flying the B787 for nearly four years since that episode — lightning-free. Did I just jinx the weather gods?

ST. ELMO'S FIRE.

Now and again, pilots will witness a static buildup on their windscreens that looks like dendritic fire strokes. Truth be told, this dancing marvel is harmless; however, it is usually a sign that nearby thunderstorms are lurking, which are not so harmless. Volcanic ash will also cause St. Elmo's fire. Passengers may see this phenomenon around propellers or near the intakes of the jet engines. Frequently, we invite flight attendants up to the flight deck to witness this vibrant light show.

HOW HIGH DO THUNDERSTORMS GET?

Most thunderstorms range from 30,000 to 60,000 feet (9144 to 18,288 m), with some topping at 65,000 feet (19,812 m), or possibly even 70,000 (21,336 m). Maximum altitudes for airliners are 39,000 to 43,000 feet (11,887 to 13,106 m) — most turboprops get as high as 25,000 feet (7620 m) — meaning we can only fly over the smaller monsters. And that's what a thunderstorm, a.k.a. cumulonimbus, is: a meteorological monster! Flying over these deathtraps, we try to give as much berth as we can. Flirting with the top of a thunderstorm is like walking gingerly on a thinly frozen lake. Small business jets can get as high as 55,000 feet (16,764 m), but they too exercise caution when flying near these nasty clouds. It's equivalent to driving into a 10-foot-deep pothole. They still make me quiver when I brush up near them. Why so close, you ask? Well, there are thousands of thunderstorms per day, with 2000 booming at any time around the globe, so their presence is inevitable. I prefer the winter season in North America over summer for that reason alone, because these badass clouds are less frequent.

→ I still get apprehensive when nearing these weather beasts, especially when I am at 30,000 feet (9144 m) and only halfway up one of these meteorological monsters. It's like a small boat coming alongside a massive oceangoing ship. Just the ripples alone could capsize the boat. Its dark and ominous presence is always respected.

→ One of the biggest fears for an airline captain is making the news because of an incident. Without a doubt, this worry enters the equation during significant turbulence and only escalates the angst. Social media is the new "six o'clock news," and no one has to be reminded that almost every passenger has a smart phone, iPad, or some other recording device.

And with onboard internet, the plane doesn't have to have landed for people to hear what has transpired. Airlines realize this and now have devoted teams to deal with issues.

Here's to smooth and pleasant flights, and as we pilots say, "Keep the blue side up!"

CAPTAIN D'S OBSERVATIONS
AND RULES TO FLY BY

- It is said a good approach makes for a good landing, but I also found a good takeoff makes for a great landing.
- Remembering emergency drills is like remembering lines in a play; all it takes is to stutter or forget one line, and watch what happens.
- Stick to the script and fly straight and level. Boring is safe.
- When meeting your flying partner for the first time, shake hands firmly, but not too firm. No one likes a "wet fish" handshake. And look them in the eye.
- When we set takeoff power, an inner smile always ensues because it means we are going flying.
- If you are captain, always offer your flying partner the chance to fly the first leg, to do the walk-around, and to buy the first beer on a layover.
- Treat everyone equally. That means the fueler, "rampies," air traffic control, and flight attendants. And don't bite the hand that feeds you, as flight attendants have means of retaliation.
- I always averred I could write a meteorology PhD thesis on the elusiveness of the seat belt sign. I turn it on, the ride gets smooth. I turn it off, it gets bumpy.

- The hardest part about my job is keeping it. This refers to the perpetual training and keeping up to date.
- Training never gets any easier. You are as good as your last simulator ride or your last landing.
- Always go the bathroom before "top of descent." Don't hold it. You never know when you may have to enter a holding pattern, perform a missed approach, fly to your alternate airport, or wait for a gate.
- All pilots love wet runways, but not too wet. Wet runways are conducive to smoother landings.
- Take pictures along the way, because the journey sure goes fast.
- You can get bad water around the world, but I have yet to have bad beer.
- When on final approach to land — remember to breathe!
- (This one is borrowed.) Keep the number of landings equal to the number of takeoffs!

✈

QUESTIONS ON THE FLY TO
TEST YOUR AVIATION IQ

Q. Most airliners around the world, except for the goliath Boeing 747 and Airbus 380, have how many jet engines?

A. You'll see two. Airbus labels them number one and number two, whereas Boeing says left and right. Incidentally, production of both the Boeing 747 and the Airbus 380 has come to an end.

Q. Who makes the most airliners?

A. Boeing and Airbus are the two main ringleaders. Embraer from Brazil and Bombardier from Canada make smaller airliners. But Bombardier succumbed to Airbus as Airbus bought the new Bombardier C-Series, now called the Airbus 220. The CRJ (Canadian Regional Jet) has been sold to Mitsubishi Heavy Industries. China, Russia, and Japan also build airliners but those remain mostly in-house.

Q. Does temperature increase or decrease with altitude?

A. Temperature decreases 2°C (3.5°F) per thousand feet (305 m). Your outside cruising temperature is a bone-chilling –57°C (–71°F).

Q. Do winds increase or decrease as you ascend?

A. They tend to increase. Some cruising winds can be over 200 knots (230 mph or 370 km/h) — stronger than any hurricane wind.

Q. Why does it take longer to fly from New York to Los Angeles than it does from Los Angeles to New York?

A. Winds tend to prevail from the west in our latitudes; they're called prevailing westerlies. The difference in flight time is about 25 minutes less with a tailwind from Los Angeles to New York.

Q. Are cargo holds pressurized and heated for animals?

A. All cargo holds are pressurized, and most are snugly warm; however, a few aircraft types keep temperatures hovering slightly above freezing. Thus there could be restrictions if you are shipping your pet.

Q. What does a large wide-body airliner cost?

Q. I'll answer this with another question: Do you have US$230 to $300 million?

Q. How fast are we barreling down the runway for takeoff?

A. We get airborne at speeds ranging from 120 knots to 175 knots (138 to 201 mph or 222 to 324 km/h). We use knots (nautical miles/hour) for speed.

Q. What altitude does an airliner cruise at?

A. Generally 28,000 feet (8,530 m) and up, with some aircraft capable of flying at just over 40,000 feet (12,192 m). Generally, the higher the better for efficient fuel burns, but usually the mid-to-high 30,000s is where most airliners soar.

Q. Is it steward/stewardess or flight attendant?

A. If you say stewardess, you will be dating yourself, or you haven't taken this Q and A. ☺

Q. How do I know what aircraft I am in?

A. Look at the safety card located in the pocket in front of you. And if you are flying on my airline, pull out the in-flight magazine to see what I wrote for the month. This magazine, as well as most in-flight magazines, gives a rundown on the aircraft fleet.

✈

GLOSSARY:
AVIATION GEEKERY
(KNOWING THE LINGO)

This glossary will help you navigate through some of the aviation jargon that stumps many passengers.

ADS-B (Automatic Dependent Surveillance-Broadcast): An umbrella of 66 low-orbiting satellites revolutionizing air traffic control. Air traffic control from above.

Affirmative: An aviator's "yes." "Negative" is "no." "Roger" is "message received" and "Wilco" means "will comply."

> ✈ A husband suspected his wife may have been getting too close to a neighbor pilot. Finally, he had enough and demanded, "Are you having an affair with that pilot?" She quickly retorted, "Negative!"

AIF (Airport Improvement Fees): Supposed to be used solely for the betterment of the airport. But it sure is a bone of contention with most passengers and airline crew traveling on employee passes. Its intent, and what ensues, varies. It is why you will always see construction at most airports. AIF is usually non-profit, meaning it is supposed to go back into the airport, which is why some airports upgrade their washrooms repeatedly. We pilots joke, "They built an airport around a construction site."

Aircrew or flight crew: Also referred to as "cabin crew" or "flight deck crew." Each airline has its own take on this, which is why it gets confusing.

Airline minute: This is a standard minute multiplied by a factor of three or four. What should be a five-minute delay translates into 15 to 20 minutes. It's like what you think an aircraft part costs. If you think a light bulb should cost $10, it really costs $30 to $40.

Air pocket: A colloquial term coined for an area of turbulence. "Pockets" in the atmosphere do not exist per se, but the term is frequently used by air traffic control, pilots, and passengers, and I occasionally partake in the colloquialism.

Alley or alleyway: Because of land constraints, or maybe due to poor planning, these narrow laneways behind the gates can cause delays when another aircraft has pushed back from the gate. You'll hear, "We are waiting for an aircraft to exit the alley before we can taxi to the gate," or, "An aircraft has pushed back into the alleyway, so we have to wait here at the gate." In confined places like LaGuardia you'll hear this a lot, but large airports like Atlanta and Denver have plenty of space. It's one reason why Hartsfield-Jackson Atlanta can move 110 million passengers annually — the most on the planet.

Altimeter: Instrument that indicates altitude of an aircraft above sea level. There is also a radio altimeter that measures height above ground at low levels.

Anti-icing fluid: Fluid that prevents ice and snow accretion and is designed to shear away during the takeoff roll. It tends to be bright neon-green.

APU (Auxiliary Power Unit): Literally a small jet engine fixed in the tail of the aircraft that supplies conditioned air and electricity on the ground. It can also supply electricity during flight. It's that hissing sound you hear when boarding or disembarking. And yes, it has the same name as the character Apu in *The Simpsons*.

Area of weather: This denotes an area of inclement weather. This may mean thunderstorms, heavy showers, or confirmed or forecast turbulence (i.e., a fasten-your-seat-belt area). As in, "Ladies and gentlemen, we are approaching an area of weather. Please return to your seats and fasten your seat belts." If it escalates or is thought to be in the moderate or heavier range, the captain will have the flight attendants secure the cabin and "strap in" as well.

"Armed and cross checked" or "Doors to arrival and cross check": The lead flight attendant will make a cabin announcement to remind the other flight attendants to either confirm the doors are armed during pushback or disarmed (chutes deactivated) approaching the gate.

ASL (Above Sea Level): Also known as AMSL (Above Mean Sea Level). Altitude of any object relative to the average sea level.

ATC (Air Traffic Control): This covers the entire infrastructure (not just the guy in the tower as depicted by Hollywood) handling everything from Tom Cruise's "flyby" to assigning holding patterns to aircraft 100 miles away. There are also area control center controllers, ground controllers, ramp (also known as apron or tarmac) controllers, and clearance delivery controllers. Some pilots will make cabin announcements using the acronym ATC, wrongly assuming passengers know the lingo.

Autoland: Most airliners have autoland capability, whereby the aircraft lands itself. It is generally used strictly in low visibilities. In fact, it *must* be used in very low visibility. The airplane, the pilot, and the landing runway must all be certified to conduct an autoland. There are no auto-takeoffs yet.

Back end: Flight attendants are known as the "back end" and pilots the "front end."

Belly: The bottom of the airplane, where your luggage is stored. Animals are stored in the belly as well but tend to be in the aft section. Sometimes you will hear their concerns from the belly.

Black box: I will borrow this definition from my friend Terry MacDonald, who wrote the book *The Black Box*. "During the 1940s, an electronic innovation was added to military aircraft. The prototype was covered in metal boxes that were painted black to prevent reflections. The 'new' electronics was referred to as the black box, an expression that made its way into postwar civil aviation and general usage."

"Bottle to throttle": A term used to depict the hours a pilot must abstain from drinking to be legal to fly. Generally, airlines require 12 hours from bottle to throttle.

Bulkhead: A dividing wall or curtain to separate sections or classes in an aircraft. Some appreciate the extra legroom a seat behind the bulkhead usually entails, but the cons are no under-seat stowage; the tray table is in the armrest; and on long-haul flights, it's where the babies' bassinets are hung. Don't be putting your feet on the bulkhead, because flight attendants will curtly ask, "Do you do that at home?" It's not classy, especially when people take off their socks.

Bumped: This means that the seats on the flight have been oversold. Sometimes a passenger may be lucky and get "bumped" up to business class, but it usually means you will be put on the next available flight. Checking in early avoids this situation. Some may volunteer to be bumped, and the rewards are high and can even be negotiated.

Cabin crew: Crew designated as the operating flight attendants.

Captain, first officer, cruise pilot, relief pilot, augment pilot: The captain is the commander, with four stripes on their epaulets and tunic sleeves, and a bit more gold embroidery on their hats. Media incorrectly refer to the captain as the "pilot." The first officer is second-in-command, with three stripes; many use the passé term "co-pilot." The cruise pilot replaces the captain or first officer for crew-rest reasons and does not land or take off. The augment pilot is usually a qualified first officer or captain and flies long-haul flights that require four pilots. Many airlines designate their cadet pilots (newly trained) with two or one stripes. The cadet pilot program is slowly creeping into North America.

"Cargo doors are closed" or "We are all closed up": This is a good sign that pushback from the gate is imminent. There are light indicators in the flight deck telling the pilots of the status of the cargo doors. We watch these doors closely as departure time nears.

Chop: Rhythmic bumps with more consistent intensity and frequency than turbulence. Equivalent to riding a bicycle on a cobblestone road or riding on the subway or train. Chop is of less concern than turbulence. It has either light or moderate intensity; there is no severe chop.

Clear, few, scattered, broken, and overcast: How cloud amounts are depicted using the fraction of 1/8s (oktas): clear is 0/8; few 1/8 to 2/8; scattered 3/8 to 4/8; broken 5/8 to 7/8; and overcast (8/8). And don't worry, "broken cloud" is not a dangerous entity like one of my passengers thought.

Cockpit: Now becoming more and more passé. See "Flight Deck."

Codeshare: An arrangement whereby one airline sells seats on a flight operated by another airline. Star Alliance and One World use codeshares. It can be a tad confusing when looking at the airport monitors, as each airline's flight number is depicted for the same flight.

Commuter: Many airline employees do not live where they work. About 30 to 50 percent of pilots commute. Less so for flight attendants. You will see these commuters next to you on a flight, either dressed in uniform or traveling incognito.

Cons: An abbreviation for contingents, not for convicts. A contingent passenger is likely an airline employee traveling on a standby, non-revenue ticket. The term refers to their seat being contingent on all revenue and senior standby ticketholders having been accommodated. "Standby" is also used.

Contrails (COTRA), condensation, or vapor trails, NOT chemtrails: Moisture from the engine exhaust freezes, causing ice-laden trails. They disperse very quickly if the air is dry; if they take tens of minutes to disperse, it indicates moisture aloft, thus weather may be nearby.

Crash pads: What pilots and flight attendants call their temporary sleeping arrangements, their home away from home when they commute. Prices vary depending on how many reside in these "unique abodes," many of which are outfitted with bunk beds. There are certainly different star ratings when it comes to crash pads.

Crew member: Person assigned to duty on an aircraft.

Cross-bleed engine start: Most airliners use compressed air from the APU (Auxiliary Power Unit) to start the engines. If the APU is not working, then a cross-bleed engine start is performed instead: an engine is started at the gate using portable external air, and then the second engine is started from cross-bleeding air. Bleed air is air taken via ducts from the compression chambers of the jet engine. It ups the pilot's load factor and is not what they want to see when showing up for work. The Dreamliner I fly uses batteries to start the engines, and we can start both engines at the same time.

Crosswinds: Takeoffs and landings are generally performed straight into the wind for maximum performance. But wind frequently blows across the runway, demanding crosswind landing and takeoff techniques.

CVR/FDR/black box: These terms frequently make the news: CVR (Cockpit Voice Recorder), FDR (Flight Data Recorder), and the infamous black box, which isn't black at all but bright orange or red.

Deadhead: A pilot or flight attendant repositioning to another airport as part of their duty: i.e., a crew member in transit. Crew may either deadhead in their uniform or in "civies."

Deicing: Removal of ice, snow, and frost accumulation on an aircraft's surface. It's the law to have these contaminants removed before takeoff. There are a few exceptions.

Direct flight: Please don't get this confused with non-stop flight. It often is. Direct flight means you are heading in the same basic direction, but it *may* mean one or several stops along the way. Think "milk run."

Domestic flights: This generally means flights within the same country; however, when a pilot says they are a "domestic" pilot, it may mean they fly within North America. There are domestic airports and international airports.

"Doors to arrival and cross check": (I repeated this one.) Another cabin announcement made by the "boss" flight attendant to their peers to confirm the doors are disarmed. Opening a door from the inside when it's armed means the chutes will explosively deploy, known as "blowing the chute." Not only is it dangerous, but it undeniably means a delay or cancellation. Yes, I've seen a blown chute, but only once in my career. When doors are opened from the outside, the chutes are automatically disarmed, except for the Boeing 737. I still get the heebie-jeebies if I must open a cabin door.

Equipment: Another name for airplane. Airline pilots can only fly the same equipment (airplane), but flight attendants can fly various types of equipment. Pilots now and again have equipment bids to select their airplane choice.

ETA (Estimated Time of Arrival): You will hear that a lot in our announcements. But is it the ETA for touchdown or at the gate? Debatable. I give it as the touchdown time because that is when passengers look at their watches to decide whether they must hustle when the cabin door opens. Add about 10 minutes of taxi time for large airports and five minutes or less for small, less busy airports.

FAA/TC: The Federal Aviation Administration and Transport Canada are pilots' friends. *Ahem.* Just like that nice police officer is your friend as you receive a speeding ticket. But seriously, they set the rules and regulations we abide by.

> ✈ Reminds me of an overused aviation meme: "Hi, I am from the FAA and I am here to help you."

Fear of Flying, a.k.a. Aerophobia: An anxiety disorder involving the sense of fear or panic some passengers experience when they fly or anticipate flying. It can be alleviated or treated by reading books like this, taking courses, or seeking professional help.

FIN or FIN number: This number is used by airlines to differentiate their fleet. The meaning of the acronym is elusive. Maybe Fuselage Identification Number, or Fleet Identification Number, or the number inscribed on the vertical tail called the fin? When a pilot calls maintenance, they address themselves using the FIN number; maintenance is not concerned about the flight number. FIN number is not to be confused with the aircraft registration, which is an alphanumeric code like that of car license plate. Aircraft registration starts with an "N" in the USA and "C" in Canada.

First class and business class: Many get this confused and think the seats in the front of the aircraft are deemed first class. There are no North American carriers with first class cabins. Some international carriers flying into North America, such as Emirates, Etihad, Singapore, Lufthansa, and Air France, have first class.

Flight attendant, NOT stewardess or steward: Saying "stewardess" instead of flight attendant reflects how little you travel or how old you are. Be cool and delete "stewardess" and "steward" from your vocabulary.

"Flight attendants, take positions for takeoff": Some form of this announcement indicates that takeoff is imminent and everyone, including the flight attendants, better be seated.

Flight crew: See "Aircrew."

Flight deck, NOT cockpit: "Cockpit" is waning, like the term "stewardess." Show people you are in the know by shying away from it. The flight deck is a pilot's office with a great view — especially from the left seat. ☺

Flight level: "Cruising level" can also be used to denote the cruising altitude. Flight levels in North America start at 18,000 feet (5486 m). Flight levels are predicated on setting the altimeter to a standard value. In the Caribbean, Europe, and most places around the world, flight levels start at unique altitudes. In Cuba, flight level starts at 4000 feet (1219 m).

Fuselage: The main body of an aircraft, and where airlines strut their livery.

Galley: Means "kitchen" in an aircraft and stems from the naval term. Stay out of the galley when meal service is happening. It's an easy way to annoy flight attendants.

Gate agent, a.k.a. ticket agent: "Ticket agent" is becoming passé in American favor of "gate agent" or "CSA" (Customer Service Agent). I know, I married one.

George: To this day, the term "George" or "George is flying" unofficially represents the autopilot system. There are two Georges on most airliners.

Go-around, missed approach, aborted landing: Sometimes a landing can't be carried out. Either a preceding aircraft hasn't exited the runway fast enough, weather is an issue, or the pilots were not set up to continue the approach, so a go-around or missed approach is performed. Yes, they can be abrupt, with some considering them aggressive maneuvers. But it's safe, albeit irregular.

Great circle: One would think a straight line is the shortest distance between two cities, but "as a crow flies" is not the shortest when talking flights over the globe. That is why your flight from London to New York would fly over the southern tip of Greenland.

"Grease it on": All pilots like smooth landings. There is nothing better for the self-esteem. If you want to stroke a pilot's ego, just tell them they really "greased it on."

Ground delay, ground stop, or gate hold: These are all delays. They're part of the traffic flow program. A ground delay is when an aircraft is delayed at its departure airport to manage demand and capacity at the arrival airport. Flights are assigned departure times, and you may find airplanes sitting at a conspicuous spot — maybe with the engines shut down to save fuel. A ground stop is a procedure requiring aircraft to remain on the ground. It may be airport-specific, related to a geographical area, or equipment-related. A gate hold is when ATC will not grant a pushback clearance, usually due to congestion.

Headwinds: Winds blowing onto the nose of the aircraft. In North American latitudes, winds generally blow from the west (prevailing westerlies), so a westbound flight would encounter a headwind. Flight planners factor this in when building flight schedules.

Heavy showers: This could be code for thunderstorms, or it may mean heavy showers. "Thunderstorm" implies very nasty weather, so many airlines tone it down.

High cloud: Classification of cloud with bases starting at 20,000 feet (6096 m). Many pilots will report high cloud in their announcements, not knowing only 2 percent of the passengers would know or care what a high cloud is.

HIRO (High Intensity Runway Operations): Air traffic control procedure to move aircraft efficiently and expeditiously. Airports are getting busier and busier, so HIRO gets airplanes moving. For landing, it's "get on, get off," and don't finesse the landing. For takeoff, it's taxi to the runway and don't dilly-dally on the takeoff roll.

Holding pattern: A racetrack-shaped course flown during weather or traffic delays. They tend to involve turns to the right, but not always. For modern airliners, it is just a push of a couple of buttons to set up a hold. But during ab initio pilot training, a student pilot must determine how to enter the hold and which way to turn. It is one of the more difficult things to nail down during a flight test. Been there!

"How's the ride?": It's what pilots ask ATC or on common frequencies, checking on flight conditions. It drives air traffic controllers nuts, as many pilots as soon as they check in say, "It's smooth" and in the same breath ask, "How's the ride?" I can hear/feel ATC cringing.

Illegal — as in "gone illegal": A crew member, either a pilot or flight attendant, who exceeds their maximum hours allowed to work per flight, day, or schedule without prone rest or a break. This could mean one crew member or the entire crew required to walk away from duty. It's a juggling constraint the airline must contend with.

In-charge (I/C), FSD (Flight Service Director), purser, lead flight attendant, or maybe "Queen Bee": The flight attendant in charge of the operating flight. Some foreign airlines called them "customer service managers" or "cabin managers."

IATA (International Air Transport Association): An organization consisting of about 290 airlines supporting aviation with global standards for safety, security, efficiency, and sustainability. The three-letter code for every airport (what you see on your ticket or bag tag) is an IATA code.

ICAO (International Civil Aviation Organization): A United Nations agency managing the administration and governance of international civil aviation. Headquarters is in Montreal. ICAO codes for airports are four letters and can be puzzling as to their origin. For example, BDA is the IATA code for Bermuda (makes sense), but how do you get TXKF for their ICAO code? Yes, it can be explained.

ILS (Instrument Landing System): Consists of the localizer and glideslope providing horizontal and vertical guidance for precision approaches. Most large airports offer ILS capability, and it's a pilot's preferred approach.

In-range checks: These checks are instigated at about 10,000 feet (3048 m), meaning about 10 minutes to landing. You may hear a double or triple chime indicating to the flight attendants that they should secure the cabin for landing.

Jet lag (circadian dysrhythmia): A physiological condition that results from alterations to the body's circadian rhythm (24-hr cycle) caused by rapid long-distance trans-meridian (east-west or west-east) travel. Generally, you don't get jet lag flying in a north-south direction.

NASA says that for each time zone traversed, one day is required to recover. That may be excessive for most travelers.

Jet stream: A narrow, meandering, fast current of air normally found at greater heights, discovered in World War II during high-altitude bombing missions. The U.S., Britain, and most of the world classify winds as "jet streams" only if they are at least 80 knots (92 mph or 148 km/h). Environment Canada classifies them as 60 knots (69 mph or 111 km/h) or more.

Jetway (bridge or jetway bridge): A bulky, boxlike, armored-looking tunnel used to connect the gate entrance to the door of the plane. Few jetways around the world have windows. Pity. This jerky, clumsily moving wheeled contraption is what breaks down when passengers want out after a 10-hour flight. By the way, it is the airport authority that owns and fixes them, not the airlines. Most jetways also supply conditioned air and power. The conditioned air can be debatable at times.

Jumpseat: An extra seat (sometimes foldable) in the flight deck for a supervisory pilot, government flight checker, training pilot, or contingent airline employee. Some larger aircraft have two jumpseats.

Jumbo jet: The mammoth four-engine Boeing 747. There are no longer any B747s (queens of the skies) flying for North American passenger airlines. The term "jumbo" never really stuck for the Airbus A380. Both the B747 and A380 are no longer being produced.

> ✈ And speaking of out-of-date terminology, how many young people know the "save" icon is a floppy disk symbol? Floppy disks are still used in some airliners to upload databases.

"Keep the blue side up": A parting farewell phrase wishing a pilot success. If you keep the blue side (sky) up and the brown side (ground) down, a successful flight will ensue.

Knot: How the aviation world measures speed. It is a nautical mile (6076 feet) per hour, not to be confused with a statute mile (5280 feet). Pilots may brag about their speed during a PA. A pilot may convert their 500-knot airspeed by doubling the value to 1000 kilometers an hour (really it is 926 kilometers per hour).

Lavatory: Airplane's name for the washroom/toilet/loo. Remember, there are smoke detectors capable of knowing when you light up. At one time, there were smoking sections on aircraft, but now if you light up, the airline's sense of humor will be flushed down the toilet.

Livery: Fancy name for paint job or airline's design.

Left seat: Physically, it is the left-side seat in the flight deck. It denotes captain, pilot-in-command, or skipper position. Peculiarly, a helicopter captain sits mainly in the right seat.

Logbook: Every aircraft must have its own logbook, which records flight legs. There is also a cabin logbook the head flight attendant fills out for unserviceable equipment, called "snags," like the seat you got that wouldn't recline. Pilots also have personal logbooks to record their flight time. Frequent flyers have them for their kids, which the pilot is asked to fill out now and again.

Long-haul flight: A flight of considerable distance and time — often with passengers suffering some significant jet lag along the way. Long-haul is about 10 to 13 hours. Ultra-long-haul is 13 hours or more.

Mach: Postulated by Austrian physicist Ernst Mach. Pronounced "mock." He divided the aircraft's true airspeed by the speed of sound. Narrow-body aircraft fly at Mach .74 to .80, whereas wide bodies fly at Mach .80 to .88.

MEL (Minimum Equipment List): A heavy book found in the flight deck, which maintenance or pilots consult to determine the serviceability of the aircraft due to a "snag." The MEL is now found on pilots' iPads.

Mile-high club: It's not about airline reward programs. Think airborne promiscuity above a mile in altitude.

Minimum connection time: The smallest amount of time allowed to change planes at an airport. If these conditions are breached, it is known as an illegal connection.

Mist: A more subdued term for fog.

Nautical mile (NM): 6076 feet. Used for distance in aviation. Not to be confused with the statute mile (SM): 5280 feet. The statute mile is used for visibility in aviation.

Non-revenue: Passenger flying free of charge (it's not really free), on a standby basis, by presenting an airline/aviation employee pass. Non-revenue passengers may or may not be on duty; therefore this expression also applies to repositioning crew members. It's also known as non-rev for short.

Non-stop: This is the flight to be on — instead of a "direct flight," as "direct" may mean one or more stops along the way.

No-shows: Passengers who either arrive late or do not arrive at all to travel on their booked flight. Gate agents are known to check the nearest airport bar for no-shows. If the no-show is traveling abroad and has checked baggage, regulations state their bags must be offloaded. Can you say "delay"?

On call: A period during which a reserve pilot or flight attendant may be assigned a flight.

Pairing: It is an aircrew's roster or schedule. Pairings change monthly. They depict flights to be flown, flight numbers, departure and arrival times, and hotel information.

PIL (Passenger Identification List): It is somewhat confidential, stating names and seat assignment and perhaps special meal requirements and frequent flyer status. Think passenger manifest.

✈ "Are you on the PIL?" is NOT a personal question.

Pointy end: It alludes to the front of the airplane. It may also reference J (business) class or first class or even the flight deck.

Pre/post-9/11: The Julian calendar includes BC and AD. In aviation, many policies and security belong to the pre- or post-9/11 eras.

Pushback: The process of moving an aircraft backwards from the gate, accomplished by coordination between the pilots and ground crew.

Ramp, apron, tarmac: Essentially the same thing, but the media likes coining it the tarmac. But they also use "tarmac" for the taxiway and sometimes the runway.

Rampie: Short for ramp attendant. A ground handling agent. You may also hear "ramp rat" tossed about. Their retort is calling pilots "glorified bus drivers."

Red alert (lightning advisory): An airport warning system when thunderstorms are within three statute miles of the airport. Outside the terminal, strobe lights may also ensue.

Red-eye: A flight that departs late at night (usually after nine p.m.) from the west and arrives early in the morning. The literal "red eye" stems from being up all night.

Reverse thrust: The temporary diversion of an aircraft engine's thrust so that it is directed forward rather than backward. This aids in slowing the aircraft upon landing. If possible, many airlines use reverse thrust sparingly to reduce wear and tear on the engines. Hence many passengers associate a nice landing with not hearing or feeling the loud rumbling caused by reverse thrust.

Right seat: Physically it is the right seat in the flight deck, but the term indicates first officer, second-in-command, or co-pilot. There are no "sideways" seats (flight engineers) anymore in modern airliners.

Seat belt extension: An addition to the regular seat belt for the more rotund passenger. They are required more and more.

Seniority: A numerical ranking system based on date of hire used by the airlines to determine positions, vacation, domiciles, monthly flights, and more. It is the pecking order for aviation careers. Some airlines try to use a fairer approach like status pay, but the seniority system rules in North America.

Simulator: Where pilots learn to fly their aircraft and to stay current on their aircraft type. They are multimillion-dollar marvels propped up on spindly hydraulic/electric jacks, capable of replicating about 500 scenarios/emergencies. You may hear it referred to as "the box," "the sin bin," or the "stimulator."

✈ Many think simulators provide the exact same experience as the airplane. In a way they do, but in another, they do not. One pilot likened it to groping a blow-up doll. It's not the real thing.

Slam clicker: A member of the flight crew who heads directly to their hotel room, "slams" the door, and "clicks" the lock. They usually stay there until crew pickup and don't socialize if they wander from their room.

Snag: An item that will need immediate or eventual fixing. Many things on an airplane do not have to be fixed right away if they adhere to the guidelines found in the MEL (Minimum Equipment List), such as a reading light, leaky water faucet or clogged drain. But the paperwork may cause a delay.

SOPs (Standard Operating Procedures): The procedures and script a pilot must adhere to according to company policy. Pilots are checked regularly on their SOP adherence. SOPs vary from airline to airline, and flight attendants have their scripts too.

Stand: Another name for gate. It's not used in North America, but abroad you will hear it.

Standby: A standby passenger holds a ticket that does not guarantee a reserved seat, meaning they are waiting for availability. "Standby passenger" may also refer to a non-revenue (contingent) passenger or employee.

> ✈ Standby could mean "stand" there and wave "bye" to the airplane. Been there while traveling on passes.

Static wicks: Stick-like devices found on winglets, the trailing edges of wings, flaps, tail, etc. to dissipate static electricity buildup.

Sterile cockpit (flight deck): To quell dangerous oversights or distractions, most airlines around the world adopted a policy whereby a pilot sticks to the script when below 10,000 feet (3048 m), for example, no extraneous chatter is allowed.

Stopover: An overnight (or possibly longer) stay at a location en route to your final destination. This is usually done to break up a very long journey, for example, London to Los Angeles with a stopover in New York.

Tail spotter: An airplane geek fascinated with aircraft/aviation. Almost every airport has them, usually found lurking along the perimeter of the premises. Many are avid photographers and some travel the world to capture airplane photos.

Tailwind: Wind in the same direction as the motion of the aircraft. A tailwind is preferred when cruising but avoided when landing or taking off.

TCAS (Traffic Alert and Collision Avoidance System): A system that interrogates other aircraft and determines if a conflict is imminent. If so, it will dictate specific immediate instructions.

"Thanks for your patience": An overused and very assuming phrase, as in, "We will be pushing back 30 minutes late because of the late-arriving aircraft. Thanks for your patience." It's akin to what is written in some super-polite emails, as in, "I hope all is well," especially when you just talked to the person in the morning.

"Top of drop": Means time of descent. Pilots will give flight attendants a "top of drop" time according to the flight management system's calculations.

Transcon: Transcontinental flights from the west coast of North America to the east or vice versa.

TSA (Transportation Security Administration): A post-9/11 agency with authority over security of the traveling public and flight crew.

Turn: As in, "I am doing a turn today," meaning a flight attendant or pilot will be flying back on the return flight during the same shift. It is usually deemed "productive flying," meaning they accumulate many flight hours in a short time. I am shocked some passengers think aircrew do turns after an eight- to 12-hour flight. Think about it.

U/S (unserviceable): "Ladies and gentlemen, it is hot in the cabin because the APU is U/S." Translation: the Auxiliary Power Unit that supplies cool air is not working.

Unruly Passenger: A disruptive passenger. There are four levels of "disruptive," and airlines take it very seriously. Some airlines will give the unruly passenger a yellow card to indicate things are getting serious. Think of it as the card given to a soccer player for unsportsmanlike conduct during a game.

Vortices (wake turbulence): Aircraft wing tips induce whirling air that can cause an abrupt bump to another aircraft, like a boat encountering waves from another boat's wake.

Wheels up time: Part of the flow control program that dictates a flight to be airborne at a certain time. A ground delay affects the wheels up time.

Wide body and narrow body: A wide-body aircraft has two aisles, whereas narrow bodies have one aisle. Some examples: the B787 is a wide body and the A320 is a narrow body airplane.

> ✈ No, "wide body" doesn't refer to the size of the pilot or flight attendant. But it's been known for pilots and/or flight attendants to "gain a few" when they regularly fly wide bodies overseas.

Winglets: Sleek appendages used to increase efficiency by reducing drag at the wing tips. Some aircraft have unique winglets, especially the newer B737s, with a dual-feather design.

Zulu, GMT (Greenwich Mean Time), or UTC (Universal Coordinated Time): "Z" for Zulu originates from the military. It's the name of the international time zone based in Greenwich, England, which all aviators reference. "Greenwich Mean Time" is passé but still frequently used, and it's not UCT for Universal Coordinated Time but UTC. Aviation is unique.

✈

INDEX

duct tape on airplane, 57–58

EEAA AirVenture Oshkosh air show, 8
earphones, 35
ears, and pressure, 100
Earth's rotation, 106
EFC (Expect Further Clearance), 77
electronic checklists, 41–42
emergency frequency, 155, 158
emergency scenarios
    in excerpt of book, 19–21
    at simulator, 18–19, 21
engines
    and altitude, 114–15
    choice of, 72
    four *vs.* two, 91, 218
    functioning, 82–83
    loss in flight test, 109
    oil, 56
    sounds, 84, 86, 87, 115
    starting of, 71–72, 75, 98
    thrust, 84
English language, 73, 115

facemasks, 141
"fam" (familiarization) flight, 11
fatigue, 178–79
fear of flying, 131–33, 191–92
female passengers, weight, 65, 66
female pilots, ratio and recruitment, 12
female voices, 120–21
filters for air in cabin, 100
final approach fix (FAF), 162, 163
Fin Number (aircraft ID), 58
firm (snug) landings, 166
first officer (F/O) (co-pilot)
    description, 38

duties, 40, 71
flying the airplane, 39–40
    at takeoff, 87–89
"Five by Five," 158
five-letter waypoints, 104–5
flatulence, 125
Fleming, Sandford, 182–83
flight attendants
    and death on board, 135–36
    job role and skills, 121–26
    public announcements, 117, 124
    schedules and hours, 43, 44
    at 10,000 feet, 92, 123
    weight, 64
flight deck. *See* deck
flight dispatch, 15
flight dispatchers, 102
flight engineers (second officers), 39
"flight hours," 43–44
flight number, 149
flight planning room, 15
flight plans, 15, 50, 102, 106
flights (or legs) for pilots, 44
flow control, 76
"flow-through" programs, 9
fluid gushing from airplane, 49–50
fog, 207–8, 212
food at airport, 156
food or drink trolleys, 125
food service, 141
four-letter codes for airports, 68–70
frequency
    for communications, 155, 161, 169
    emergency frequency, 155, 158
    and PAs, 119–20
frontal turbulence, 197
frost, 212

fuel
in cold weather, 211, 212
jettison ("fuel dumping"), 160
quantity, 49, 56
saving methods, 65
fuselage, names on, 63

Gander (Newfoundland), 1
gate (airport gate), 51, 156
gate agents, 128
Gimli (Manitoba), and pilot license of
author, 6, 16–17
glide slope, 146, 150–51
"glory" and "glory of the pilot," 203
glossary, 221–42
GMT (Greenwich Mean Time), 182,
183
GNE (Gross Navigational Errors), 106
go-around, 164
"going off the rails" ("rolling delays"),
54
GPS (Global Positioning System), 104
"great circles," 103
Greenwich (UK), 184
ground-based navigation devices, 104
ground delay program, 76
Ground Power Unit (GPU), 71
ground stop, 76, 77
gushing fluid from airplane, 49–50
gust suppression system, 195

Halifax Flying Club, 6
hard landings, 166–67
hats, 14
headings (holding patterns), 76, 77
headwind and tailwind, 57
heat and flying, 208–9

Heathrow (London), security for
aircrew, 32–33
"heavy" aircrafts, 159–60
helicopter pilots, 9
HEPA (High Efficiency Particulate Air)
filters, 100
HF (High Frequency) radios, 158
Hiltz, Clement, 18
hiring by airlines, 9
HIRO (High Intensity Runway
Operation) mode, 164, 167
holding patterns (headings), 76, 77
Hopkins, Carolyn, 29–30
hours and schedules, 43–44
Howard, Luke, 186–87
HUD (Head-Up Display), 88
human trafficking, 138–39
humidity, 206–7
hydration with water, 98, 100

IATA (International Air Transport
Association), 54, 67
ICAO (International Civil Aviation
Organization), codes for airports,
68–69
Icarus, 188
identification for bags, 34
ILS (Instrument Landing System),
145–47, 150
"in" and "out" times, 51, 66
information kiosks at airport, 157
instrument approaches, 145–46
instrument flight rules (IFR), 151
interior design of airplanes, 151
international phonetic alphabet, 159
internet, 190
iPads, 50

winter

    extreme cold, 210, 211

    flights facts, 108–9

    and weight, 64, 66

wireless devices, 152–53

World War II bombing mission, 199

"Y" codes, 67–68

yellow line, 74

YYZ (Toronto airport/Toronto
    Pearson)

    code for, 67–68

    deicing, 204

    noise abatement, 83

    takeoff process, 87–89

    waypoints, 105

Z or Zulu, 182

Printed on Rolland Enviro.
This paper contains 100% post-consumer fiber,
is manufactured using renewable energy - Biogas
and processed chlorine free.

100%

PCF

BIO GAS
ENERGY

PERMANENT

	MIX
FSC	Paper from responsible sources
www.fsc.org	FSC® C103567